Flying in Adverse Conditions

Other books by R. RANDALL PADFIELD

Learning to Fly Helicopters

Cross-Country Flying—3rd Edition
previous editions by Paul Garrison and Norval Kennedy

To Fly Like a Bird

Other books in the TAB PRACTICAL FLYING SERIES

Multiengine Flying *by Paul A. Craig*

Pilot's Avionics Survival Guide
by Edward R. Maher

The Pilot's Air Traffic Control Handbook—2nd Edition
by Paul E. Illman

Advanced Aircraft Systems *by David Lombardo*

Stalls & Spins *by Paul A. Craig*

The Pilot's Radio Communications Handbook—4th Edition
by Paul E. Illman

Night Flying *by Richard F. Haines and Courtney L. Flatau*

Bush Flying *by Steven Levi and Jim O'Meara*

Understanding Aeronautical Charts *by Terry T. Lankford*

Aviator's Guide to Navigation—2nd Edition *by Donal Frazier*

ABCs of Safe Flying—3rd Edition *by David Frazier*

Be a Better Pilot: Making the Right Decisions *by Paul A. Craig*

Art of Instrument Flying—2nd Edition *by J.R. Williams*

Good Takeoffs and Good Landings—2nd Edition
by Joe Christy Revised and Updated by Ken George

The Aviator's Guide to Flight Planning *by Donald J. Clausing*

Avoiding Common Pilot Errors: An Air Traffic Controller's View
by John Stewart

Flying In Congested Airspace: A Private Pilot's Guide *by Kevin Garrison*

General Aviation Law *by Jerry A. Eichenberger*

Improve Your Flying Skills: Tips from a Pro *by Donald J. Clausing*

Flying in Adverse Conditions

R. Randall Padfield

Illustrated by
Ralph C. Padfield

TAB Books
Division of McGraw-Hill, Inc.
New York San Francisco Washington, D.C. Auckland Bogotá
Caracas Lisbon London Madrid Mexico City Milan
Montreal New Delhi San Juan Singapore
Sydney Tokyo Toronto

Published by TAB Books.
TAB Books is a division of McGraw-Hill, Inc.

pbk 4 5 6 7 89 10 DOC/DOC 9 9 8
hc 2 3 4 5 6 7 8 9 10 DOC/DOC 9 9 8 7 6 5

Library of Congress Cataloging-in-Publication Data

Padfield, R. Randall.
Flying in adverse conditions / by R. Randall Padfield : illustrations by Ralph C. Padfield.
p. cm.
Includes index.
ISBN 0-07-048139-3 ISBN 0-07-048140-7 (pbk.)
1. Airplanes—Piloting—Safety measures. 2. Meteorology in aeronautics. I. Title.
TL710.P33 1994 94-7704
629.132'5214—dc20 CIP

Acquisitions editor: Jeff Worsinger
Production team: Katherine Brown, Director
Jan Fisher, Desktop operator
Susan E. Hansford, Typesetting
Stephanie Myers, Computer artist
Elizabeth J. Akers, Indexer
Design team: Jaclyn J. Boone, Designer
Brian Allison, Associate Designer
Cover photograph: Douglas Robson PFS
Back cover copy: Cathy Mentzer 0481407

To William C. and Violet Dettmann, my grandparents.

Contents

Acknowledgments

MANY PEOPLE CONTRIBUTED TO THE BOOK THAT YOU HOLD IN YOUR hands. Among them, Ralph C. Padfield, my father, drew all of the line drawings in *Flying in Adverse Conditions*, as he did for *Learning to Fly Helicopters*, which was my second book for TAB/McGraw-Hill. His experience as a Navy pilot during World War II, and his years as an engineer for Bethlehem Steel Research, provided him with both an aviation and technical background. His strong desire for accuracy and detail in everything he does ensures that his carefully drawn illustrations are not only beautifully done, but technically accurate. He also contributed one of the hangar stories.

The technical editor for the book was Mark Phelps, an accomplished pilot, an experienced writer and editor with *Aviation International News*, and a good friend and colleague. He read the entire manuscript for both editorial and technical accuracy, making many, many valuable suggestions and recommendations, including changes to the arrangement of some chapters. I cursed him mightily several times because I knew that he was right, and eventually I figured out how to accomplish most of his suggested improvements. Thanks to him, the book is much better than it would have been without his much-valued input.

Eileen Clark, production editor at *Aviation International News* and another of my colleagues, prepared the fine charts found in appendix B. Like Mark, she worked on these during her free time—what little of it there is—using her considerable expertise in computer graphics and layout. Her dedication to accuracy and detail rivals that of my father.

The foreword was graciously written by Bill Garvey, now a senior editor with *Reader's Digest*, and formerly the editor of *Flying* and *Professional Pilot* magazines. One of the most respected editors in aviation writing, Bill is also a very fine writer. If you ever see his byline on an article, you know the article will be interesting, accurate, and well written. Bill published my first commercial article in *Professional Pilot* back in January 1981, and I was honored that he agreed to write the foreword for *Flying in Adverse Conditions*.

Many pilots have told me how much they really like and learn from real-life flying stories, and I am sincerely grateful to the pilots who have allowed me to include their experiences in this book. Larry Bothe sent me five good stories and I used them all. Other stories were contributed by Dean Borghi, Dorothy Regan, John Schubert, and Brian Stoltfus. Some of the pilots who contributed hangar stories wish to remain anonymous and I have honored their request.

A special thanks to Martin Leeuwis for permission to use many cartoons from his book, *Say Again*. Humor is often a good way to make a serious point and Martin's cartoons always hit the mark. Martin has also published another book of cartoons, *Say No More*, and I highly recommend both books for those of you who would like to add some humor to your aviation library. You may obtain them from Martin Leeuwis Publications, P.O. Box 580, 3740 AN Baarn, Holland.

Another special thank you to Drew Steketee, senior vice president of communications at the Aircraft Owners and Pilots Association (AOPA), and Roberta Toth, manager of AOPA's library/research section, for much help obtaining background material used in writing this book.

I also would like to acknowledge the following organizations for permitting me to use their photographs: Allied Signal Incorporated, *The Bear Valley Voice*, B.F. Goodrich Flight Systems, Cessna Aircraft Company, Lockheed Aircraft Service Corporation, Pilatus Aircraft Limited, Piper Aircraft Corporation, Stoddard-Hamilton Aircraft, and Sony Electronics, Incorporated.

Saving the best for last, I am eternally grateful for the constant, unwavering support and love from my wife, Moira, and our children, Heitha, Dirk, and Thomas. Writing a book requires sacrifices from the whole family, and they have all been more than understanding. Now that this book is finished, we can return again to a normal home life...at least until the next time the muse comes knocking at my door.

Foreword

WHEN I WAS LEARNING TO FLY, MY MOST TREACHEROUS ADVERSARY became apparent on my fifth or sixth excursion aloft: engine screaming, stomach sinking, yoke jumping, stall warning horn blaring, and then an unexpected, terrifying plummet to earth. My instructor called it the power-on stall. I almost called it quits.

But I didn't, and eventually I came to master and even enjoy what was, after all, simply a maneuver.

It was on my very first solo cross-country that I glimpsed a true menace. It snuck up on me. I was flying low across Lake Okeechobee on a dull, overcast afternoon. As the shoreline disappeared behind me, the gray water merged with the gray horizon, which merged with the gray sky. Quite suddenly all my body's sensors short-circuited, simultaneously insisting that I was turning turtle, clawing heavenward, and driving straight for the center of the earth. My instant panic subsided just as quickly when I looked back and saw farmland behind and below, thus resolving the debate and confusion within.

The brief episode taught me a vital lesson: stay alert to weather, even the most benign, because it is second only to bad judgment as the pilot's most dangerous foe.

The best way to learn about weather is through experience, and the experience of others should be your primary reference because it is the richest and most varied. Put another way, you don't want to learn some things firsthand.

That you have selected this book is clear evidence that you understand the value of others' experience. What you might not yet appreciate—but will as you read on—is just how well you have chosen your weather adviser.

Most of us strive to do one thing well, professionally. But there is a special group of people who strive for and succeed at more: Randy Padfield is a member of that group.

First, he is a command pilot with extraordinary flight experience. To survive as an Air Force search and rescue pilot in Iceland and Alaska, as an offshore helicopter pilot in the North Sea, and as a commuter/corporate pilot in the northeastern area of the United States, you need consummate piloting skills, an iron will, and a complete understanding of your abilities, your craft, and your environment.

Secondly, he is a gifted writer able to convey his knowledge, his experience, and the experience of fellow airmen with refreshing clarity and feeling.

Take advantage of these experiences and insights in *Flying in Adverse Conditions*, and weather—the most changeable and formidable of adversaries—won't take advantage of you.

WILLIAM GARVEY
Former editor-in-chief,
FLYING and *Professional Pilot* magazines

Introduction

FLYING HAS NEVER BEEN FOR THE FAINT-HEARTED. DAEDALUS AND ICARUS, Orville and Wilbur, Amelia, Igor, Lindy and Anne, Beryl, Antoine, and all the other early pilots were many things—dreamers, visionaries, geniuses—but they weren't cowards. To be a pilot during the pioneering days of aviation took courage, guts, moxie, whatever you want to call it. Even the pioneering passengers had to be courageous.

Things have changed. Flying is safer today than it was during those thrilling days of yesteryear, much safer. Millions of people board airplanes every year with as much concern as if they were boarding a downtown bus; depending upon the city, they probably board airplanes with even *less* concern than they board downtown buses.

Statistics prove that flying is the safest form of travel. Seniors, young couples with children, business travelers, and babies with their eyes all aglow are all willing air travelers. And yet...things have not changed that much.

To be a *pilot* still takes guts.

Flying might not be nearly as life-threatening today as it was when Lindbergh crossed the Atlantic or Byrd flew to the North Pole or de Saint-Exupery flew over the Sahara, but the weather and surface conditions that increased the dangers of their history-making flights are still with us. Mankind has improved the flying machine and aviation's host of supporting services, things like weather reporting, flight following, navigation, and search-and-rescue, but nature can still be as cunning and unforgiving as ever.

Even if you make it a habit to fly on only windless, cloudless, picture-perfect days, and never plan to cross an ocean or fly north to the Arctic, it's likely you'll encounter some kinds of adverse flying conditions if you fly often enough. All flights are not created equal; some flights are simply more dangerous than others.

This book is about those more-dangerous flights. It's about the things that make your heart beat a little faster and your palms start to sweat. It's about flying in non-CAVU weather conditions and over hostile terrain.

Non-CAVU flying? Why would any sane person want to fly when the ceiling and visibility aren't unlimited, or the landscape below is hazardous? Why pit one's luck against the awesome, impersonal forces of nature?

There are many reasons actually. Sometimes the shortest way from A to B is through some nasty weather or over a stretch of hostile terrain. Sometimes the *only* way from A to B is through some nasty weather or over a stretch of hostile terrain. Sometimes, to quote the poet Robbie Burns, even "the best laid schemes o' mice and men gang aft a-gley," and when they do, we can get caught up the proverbial creek without a paddle.

This book provides you with the information you'll need to avoid adverse conditions whenever you can and gives you ways to cope with them when you can't. It will give you a few spare paddles in your canoe when you get caught up that creek.

Flying in Adverse Conditions is divided into two parts. The first part covers the typical sequence of a flight, from planning to shutdown, similar to the format found in chapter 5 of the *Airman's Information Manual*. (Indeed, this comprehensive publication from the Federal Aviation Administration is referenced frequently, not only because of the good information found there, but as an incentive to pilots to spend more time with it.)

Part 2 covers the many adverse conditions that a pilot is apt to encounter in flight, from weather and terrestrial hazards, to the problems of human factors. Tips about night flying and how to handle emergencies are also included.

The chronological nature of Part 1 presented some challenges. Thunderstorms, as an example, must be considered during all phases of the flight, from planning to landing. How should storms and other adverse hazards be handled? Two ways.

In Part 1, hazards are discussed with respect to their impact on a flight, but not in detail. For example, during the planning stage, a convective sigmet, which is a report of thunderstorms for a particular area, might cause you to postpone, reroute, or even cancel a flight. In Part 2, thunderstorms are discussed in much more detail, including tips on how to avoid them in flight and what to do if they can't be avoided.

Throughout the book you'll find hangar stories. These are real-life accounts of actual incidents that occurred in flight, stories that you might hear while hanging around a hangar on a foggy day, at a fly-in breakfast, or anywhere pilots gather. Some of the stories are my own, but most are from other pilots. All of the stories have a point relating to that chapter's topic. Hopefully you'll find the stories entertaining and enlightening.

I also encourage you to take a look at appendix B. It presents a summary of the latest statistics from the FAA about weather-related accidents. I think you'll find the information that it contains intriguing.

A final word about philosophy. I believe that experience is a good teacher, but not always the best teacher. From the beginning of my 20-plus years in aviation as a military and a civilian pilot, I accepted that it's better to learn from the mistakes of others than to make those same mistakes myself. There is much a person must individually learn and experience in aviation, but there is even more to learn from the experiences of those who have gone before.

Because you have chosen to read this book, I assume you share the same philosophy, and I'm glad you do. I sincerely hope that *Flying in Adverse Conditions* adds to your knowledge of aviation and weather and helps you continue to be the safe, mature pilot that you obviously already are.

R. RANDALL PADFIELD
Bethlehem, Pennsylvania
January 1994

Alan Stats

"Aviation in itself is not inherently dangerous. But to an even greater degree than the sea, it is terribly unforgiving of any carelessness, incapacity, or neglect." (Origin unknown.)

1
Preflight planning

"I'd much rather be down here, wishing I were up there, than up there wishing I were down here."

Painted on the side of a U.S. Navy jet.

THERE'S AN OLD SAYING THAT SAYS YOU CAN HARDLY EVER FLY OVER 500 miles without a change in weather. Sometimes the weather changes before you even leave the traffic pattern. The longer the flight, the greater the chance that something unexpected will occur. Good planning and preflighting are the best ways to prepare for expected adverse conditions. They also are the only ways to prepare for unexpected adverse conditions. Indeed, the *Airman's Information Manual* lists "inadequate preflight preparation and/or planning" as the most frequent cause for general aviation accidents that involve the pilot in command.

REGULATORY IMPORTANCE

FAR 91.103 makes the FAA's standpoint on the importance of preflight action very clear. It states in part: Each pilot in command shall, before beginning a flight, become familiar with all available information concerning that flight. This information must include—(a) For a flight under IFR or a flight not in the vicinity of an airport, weather reports and forecasts, fuel requirements, alternatives available if the planned flight cannot be completed, and any known traffic delays of which the pilot in command has been advised by ATC; (b) For any flight, runway lengths at airports of intended use, and...the takeoff and landing distance data contained in the approved aircraft flight manual.

Lest there be any doubt about the weight of the above regulation, litigation lawyers have successfully used it in court to prove the negligence of pilots who failed to "become familiar with all available information concerning the flight." Granted, in reality, this is an impossible task, but a pilot stands on much firmer legal ground if at least an attempt is made to obtain "all available information" than if nothing at all is done.

The thoroughness with which preflight preparation is conducted depends to a considerable degree on the amount of experience of the pilot, the type of airplane to be flown, and whether passengers unfamiliar with light-aircraft travel will be carried. Even if an experienced pilot is very familiar with the destination and the route, a quick check of the weather and NOTAMs before taking off is about the lowest cost insurance that can be "bought."

Waiting to check weather or talk to a flight service station specialist until at cruising altitude is unprofessional for pilots of every experience level. At the very least, you might save yourself time and fuel by checking the weather before zooming off into the wild blue yonder. At best, you might escape taxing your capabilities to the limit because you encounter a few weather surprises or you might avoid inadvertently flying into an active restricted area because you didn't know it was "hot."

PRESTEPS: BEFORE YOU START

The standard preflight steps include checking the weather, filing a flight plan, and taking a look at the aircraft; however, from my experience as a military and professional commercial pilot, I learned that a safe flight begins long before I walk into the flight office an hour or two before a scheduled departure.

Table 1-1. Preflight steps

Presteps: Before you start
Prestep 1. Are you fit to fly?
Prestep 2. Check the "big weather picture."
Preflight steps
Step 1. Plan and plot the route.
Step 2. Fill out a navigation or flight planning log.
Step 3. Check the weather.
Step 4. Call the intermediate and destination airports.

Step 5. Check the weight and balance and performance charts.
Step 6. File a flight plan.
Step 7. Preflight the airplane.
Step 8. Brief your passengers.
Step 9. Organize the cockpit.
Step 10. Start the engine and go!

Prestep 1. Are you fit to fly?

Often overlooked by private pilots, and sometimes disregarded by commercial pilots, is their personal condition when they get in the cockpit. The fact that the vast majority of accidents—something like 80 percent—are due in part to human errors makes this regrettable.

The day before any flight you should be mentally and physically preparing yourself. This preparation relates to such things as sufficient rest, nutritious meals, no unauthorized medications, and at least eight hours from "bottle to throttle"—no alcohol at least 8 hours prior to flight. (Chapter 8 covers these subjects in more detail.)

It also relates to your currency as a pilot. Is your medical certificate current? Have you had a biennial flight review or the equivalent in the last 24 months? If you will be flying passengers, have you made three takeoffs and landings in an aircraft of the same category and class as the one you'll be flying? If you'll be flying at night with passengers, have you made three takeoffs and landings after sunset to a full stop in the same category and class of aircraft? If you're instrument-rated and plan to file IFR, have you flown 6 hours of actual or simulated instruments and six approaches during the last six months?

If you are not current, you should not attempt the flight. To do so is to tempt fate first, and the wrath of the FAA afterward.

Prestep 2. Check the "big weather picture"

Most pilots have a greater-than-normal interest in weather because it's such an important part of aviation. After all, no matter where you fly, you always end up occupying the same airspace as some kind of weather—unless you're in orbit.

The day before a flight is a good time to take a look at the weather. Your goal is to develop an awareness of the "big picture." "A.M. Weather" on public television stations, "The Weather Channel" on cable TV, newspaper weather maps, and TV and radio weather reports are all good sources for preliminary flight planning.

I find "The Weather Channel" particularly good for planning 24 hours ahead, sometimes even longer. Although not geared specifically for aviation, TWC gives you a very good idea of the major weather systems that might affect your route of flight and your current local conditions. Watching the channel for a half hour every day is almost like taking a course in weather forecasting. The local radar summaries will also help you relate precipitation intensity level reports to what the actual conditions are like.

When heavy weather comes through my area, I like to turn on "The Weather Channel" to see what the radar summary looks like—and to check how accurate it is. Most of the time it's very accurate.

When I was flying Air Force rescue helicopters in Alaska, I made it a habit to watch the "A.M. Weather" broadcast before going to work. If I was scheduled to fly that day, it gave me a good idea of what to expect and often included some information that we didn't get from our Air Force weather briefers, who concentrated more on the higher altitudes for the F-4 and C-141 drivers. Of particular importance to us were the numerous mountain passes that we had to fly through. Because "AM Weather" was geared more toward the many small airplane pilots in Alaska, it gave us the information we needed to fly safely at the lower altitudes, too.

After checking the "big weather picture," you might decide to cancel or delay your flight, or select another destination. If the weather is borderline and you want more information, go ahead and give flight service a call now. Often it's easier to get a hold of a briefer during the less-hectic evening hours. If you don't call flight service, at least you'll have a general idea of the weather conditions to expect and you can plan accordingly.

Okay, you're rested, medically safe, and current, and the weather forecast seems reasonable. Let's get on with the preflight planning.

PREFLIGHTING STEPS

No hard and fast rules say that you must prepare for your trip in the following order, or that you even must include all the steps. But if you do each of the steps thoroughly, you'll be "familiar with all (or nearly all) available information concerning the flight" and you'll be as prepared as you can be for a flight in any condition.

The order of the first three steps is open to discussion, maybe even the fourth step, too. You might, for example, have no particular destination in mind other than an airport that has a decent restaurant nearby (Fig. 1-1). In this case, Step 4, a call to the local fixed-base operator, could be the first thing you do. After you inquire about the dining possibilities, you can ask about weather and other special conditions.

Or, you might have a particular city in mind as your destination, but have a choice of airports. A few calls to find out such things as tie-down rates, the availability of hangar space, and ground transportation will help choose the airport.

When you decide on your destination airport, the main questions are, "Is the weather good enough to get there?" and "What's the best route?" The answers to these two questions are often related. Sometimes you might have to repeat Steps 1 through 4 several times before you find the route, weather, and destination airport that works best.

Step 1. Plan and plot the route

It's often convenient, if your route requires several sectional charts, to take a look at a chart that covers a larger area than a single sectional, such as a world aeronautical chart or a National Ocean Service flightcase planning chart.

Fig. 1-1. *Your destination might be more important on some flights than others. A breakfast fly-in at a small airfield attracts many flyers when the weather is good, but when adverse weather sets it, it's often wise to pick another destination for your weekend jaunt—or even postpone any flying until the weather improves.*

Spread the current sectional charts for your flight on a large flat surface. Using a nav plotter or ruler and a pen or pencil, draw a line on the route to be flown, being careful to check for restricted and prohibited areas and other airspace that might have to be avoided. If you're planning a long flight, you'll need to figure refueling stops. Consider your passengers: Will they need a break on the ground after 2 or 3 hours in flight? Your airplane might have 6 hours of endurance, but few people do.

This is the time to look for alternate routes and even alternate destinations. Even though the FARs might not require a flight conducted under visual flight rules to have an alternate, it's not a bad idea to always have one. What do you do, for example, if another aircraft crashes at a single-runway airport? The airport could be closed to all other traffic for several hours or even longer. It has happened. Look for airports near your route of flight that could be used as precautionary landing areas.

There's one last bit of information you'll need that unfortunately you can't get directly from a chart: the traffic pattern altitude at the airport. Adding 800 or 1,000 feet to the airport elevation yields an estimated pattern altitude; specific altitude might be determined by size and weight of an aircraft. You have to look this up in the *Airport/Facility Directory* or *AOPA's Aviation USA*, which is all right because you need the telephone number of an FBO at the airport anyway for Step 4.

Step 2. Fill out a navigation or flight planning log

This might seem unnecessary to you, particularly if you've been over the same route before and have sophisticated navigation equipment in your airplane, like loran or GPS. Perhaps you never use the flight log for anything except to enter the route into the navigator, if the airplane is so equipped. On the other hand, you never know when the electronic navigator might decide to take a vacation and you find yourself staring at a blank screen. No problem because you have a sectional? Let's add some unforecasted low clouds or fog or haze or thunderstorms and then the situation becomes more critical. With a completed flight log handy, it's much easier to keep track of your position.

Numerous flight planning forms are available that present more or less the same information in different formats. Find one you like and buy it in bulk. Or devise your own and make copies of it. In a pinch, you can use a blank sheet of paper; lined paper is preferable when you're trying to read something quickly in the cockpit.

It's also smart to write down the various frequencies you'll need during the flight. ATIS and AWOS/ASOS frequencies for airports you plan to over-fly will make it easy to obtain the latest weather, including the local altimeter setting, which is important to maintain the proper altitude.

Step 3. Check the Weather

Experienced pilots obtain weather information in three steps. The first step is to get the big picture, which you did in the prestep. Now you need to obtain a more detailed analysis of the weather. There are numerous ways to do this without even talking to a meteorologist or flight service station weather briefer. You can call a free automated weather information service, such as 1-800-WX-BRIEF, or one you pay for. If you have a computer and a modem, you can use the Direct User Access Terminal (DUAT) system. You can use one of the commercial computerized weather services, often available at FBOs. Certain vendors also provide weather information via facsimile.

Whichever method you use to obtain a detailed weather briefing, the goal should be to find out the following information:

- Adverse conditions: Any significant meteorological and aeronautical information that might influence your proposed route of flight, or even prompt you to cancel the flight entirely.
- Synopsis: A brief explanation of the general weather (fronts or pressure systems) that might affect the flight.
- Current conditions: A summary of the current weather at the departure point, destination, and points in between. Sequence weather reports and current pilot reports are the most common sources of current conditions.
- En route forecast: A summary of forecast conditions along the route. This information is available from several automated sources: terminal forecasts at airports along the route, 800-WX-BRIEF, DUAT, and the like.
- Destination forecast: The forecasted weather, including any significant changes, for the destination at the planned time of arrival, plus and minus one hour. Terminal forecasts provide this information.

- Winds aloft: Forecast and observed winds and temperatures at desired altitudes along the route. Use the standard temperature lapse rate (2°C per 1,000 feet) to help determine the freezing level, but be aware that temperature inversions are common in many parts of the country.
- Notices to airmen: Relevant NOTAMs to the proposed route of flight.

With the above information in hand, you can usually make your go-no-go decision.

If the weather still seems iffy, take the third step and contact an FSS briefer in person or by telephone. Even the most sophisticated automated systems can't be compared to a human briefer. Generally speaking, the longer the flight or the more changeable the weather, the greater the chance you'll need to talk to a briefer. Often the briefer will be able to provide more timely and in-depth information than is available from the automated systems and will usually give a quite insightful analysis of the current weather situation, too.

If the briefer tells you "VFR flight not recommended," don't ignore this advice. This admonition must be given when there are conditions along your route of flight that are strictly IMC or are marginal VMC. If it's marginal, you still might be able to make the flight VFR depending on your knowledge of the route, your skill level, your aircraft and equipment, the local area, and how bad the weather really is. If it's definitely IMC, you shouldn't go unless you are able and legal to file IFR. Ask the briefer why the admonition is being made and then consider if you are equipped to handle the situation.

Forecast reliability

Although meteorologists understand some atmospheric behaviors, they have watched the weather long enough to know that their knowledge of the atmosphere is far from complete. The wise pilot continually views aviation weather forecasts with an open mind. She knows that weather is always changing and that the older the forecast, the greater the chance that some part of it will be wrong. To have complete faith in forecasts is almost as bad as having no faith at all.

Studies of aviation forecasts have indicated the following:

- A forecast of good weather (ceiling 3,000 feet or more and visibility 3 miles or greater) is much more likely to be correct than a forecast of conditions below 1,000 feet or below 1 mile.
- If poor weather is forecast to occur within 3–4 hours, the probability of occurrence is better than 80 percent.
- Forecasts of poor flying conditions during the first few hours of the forecast period are most reliable when there is a distinct weather system, such as a front, a trough, or precipitation. There is a general tendency to forecast too little bad weather in such circumstances.
- The weather associated with fast-moving cold fronts and squall lines is the most difficult to forecast accurately.

- Errors occur when attempts are made to forecast a specific time that bad weather will occur. Errors are made less frequently when forecasting that bad weather will occur during some period of time.
- Surface visibility is more difficult to forecast than ceiling height. On the other hand, actual ceiling height is often more difficult to measure than visibility. According to ICAO regulations (but not FARs), a pilot may start an instrument approach to an airport if the visibility is reported above minimums; reported ceiling height need not be considered.
- Visibility in snow is the most difficult of all visibility forecasts.

Forecasters *can* predict the following at least 75 percent of the time:

- The passage of fast-moving cold fronts and squall lines within ±2 hours with as much as 10 hours in advance.
- The passage of warm fronts and slow-moving cold fronts within ±5 hours, up to 12 hours in advance.
- The rapid lowering of the ceiling below 1,000 feet in prewarm front conditions within ±200 feet and within ±4 hours.
- The onset of a thunderstorm 1–2 hours in advance, if radar is available. This is improving considerably with the implementation of the next-generation weather radars (NEXRAD).
- The time rain or snow will begin, within ±5 hours.

Forecasters *cannot* predict the following with an accuracy that satisfies present aviation operational requirements, although newer technologies, like NEXRAD and wind profilers, are helping forecasters improve these numbers:

- The time freezing rain will begin.
- The location and occurrence of severe or extreme turbulence.
- The location and occurrence of heavy icing.
- The location of the initial occurrence of a tornado.
- Ceilings of 100 feet or less before they exist.
- The onset of a thunderstorm that has not yet formed.
- The position of a hurricane's eye to nearer than 80 miles more than 24 hours in advance.

Hangar story

Some years ago, at a Norwegian Air Force station in Bodø, Norway, the pilots became disenchanted with the local meteorologist, whose forecasts, they complained, were more often wrong than right. They thought they could do better with their own predictions and devised a simple formula. They predicted that tomorrow's weather would be the same as today's.

The climate on Norway's West Coast probably has some of the most changeable weather in the world, so this was a rather risky way to make a forecast. But because most weather systems usually take more than a day to transverse the area, the pilots figured they had a fairly good chance of being correct.

Of course, when the weather did change, their forecast for that particular day was wrong, but then they would simply modify their forecast for the next day. Over a period of several months, the pilots ended up being right better than 50 percent of the time. Compared to the meteorologist, they had a better record.

The point of this story is not to suggest that you ignore the forecasts of meteorologists and blindly believe that tomorrow's weather will be the same as today's. The point is to encourage you to develop your own weather awareness that you can supplement with the information you obtain from forecasters.

Realize that weather forecasting is fraught with so many variables that it's amazing meteorologists are ever right. Local topographical and man-made features can also create weather in small areas that is completely different from weather that is observed and forecast at an airport less than a mile away.

Get a good weather briefing, but don't leave your personal knowledge of the local area, and your common sense, behind.

Definite no-gos

Only a few weather conditions should cause you to delay a flight, if not cancel it. If these no-go conditions are merely forecast and not yet observed, you might still be able to make the flight, but do so with caution and with continual updates on actual conditions.

Sigmets and airmets in the forecast. Sigmets are advisories of potentially hazardous weather conditions of concern to all aircraft, such as icing, turbulence, dust storms, sandstorms, and the like, and are issued as necessary. Airmets are advisories of potentially hazardous conditions, mainly of concern to small aircraft and are issued as necessary except when already part of an area forecast. Airmets concern weather of less severity than sigmets, but if the conditions actually materialize, they're serious enough to cancel most flights in small aircraft.

Convective sigmets. These advisories of concern to all aircraft are issued hourly during periods of hazardous convective weather, such as tornadoes, thunderstorms, and hail. They are issued by the National Severe Storms Forecast Center in Kansas City. Convective sigmets should be taken very seriously. Because convective weather often passes through an area quickly and the convective sigmets are updated hourly, it might be possible to simply delay a flight for a few hours, rather than cancel it completely.

Icing. Icing is often difficult to forecast, but some conditions just about guarantee it. Freezing rain is one of them, although accurately predicting the time when freezing rain will begin is almost impossible. When the surface temperature is just above freezing and low clouds and precipitation are present, you can usually expect icing in flight. The worst icing conditions occur within a few degrees of freezing. Even aircraft with deice and anti-ice equipment have a hard time in severe icing. If your airplane has neither, don't even try it.

Small temperature/dewpoint spread. This is easy to overlook in a forecast, but can be very significant. When the temperature and dewpoint are within 4° and relative humidity is high, conditions are ripe to create fog. If you'll be flying over terrain that increases in elevation, there's a good chance the ground will rise up to meet the cloud layer. If the overall temperature is dropping, too, as often happens as the sun goes down, the ceiling will come down as well. Don't get caught with rising terrain in front of you and lowering temperatures behind you.

High winds. Check the winds along your route of flight. Perhaps it's calm at your departure and destination airports, but if it's blowing 30 knots near the mountain pass you must fly through, it might be too turbulent for your aircraft. Remember that surface winds are often less than the wind at 1,000 or 2,000 feet, so if it's 30 knots on the surface, it might be 35 knots, 45 knots, or even more at your intended flight altitude. Check the winds aloft forecast. You might get lucky and find a band of relatively calm wind at a usable altitude. But be wary: Winds at altitude are very changeable and forecasts are more often wrong than right. It's hard to give a rule of thumb about wind, but if the reported surface wind anywhere along your route of flight is in excess of twice the demonstrated crosswind component of your airplane, you have a good reason for not making the flight.

Frontal systems. If your route goes through a front, there's a good chance you're in for some adverse weather. Some fronts are worse than others, of course, but be wary. Studies of aviation forecasts have indicated that the weather associated with fast-moving cold fronts and squall lines is the most difficult to forecast accurately. If along the route the forecasts indicate large variations in wind direction, wind velocity, ceiling heights, and visibilities, expect the worst.

Sunset. If the day is drawing to a close and it's not clear-and-a-million over the planned route, don't try it. It's tough enough flying IFR at night in low-visibility conditions, but at least at IFR minimum altitudes terrain clearance is assured. VFR at night is hard enough when the weather is clear; add some haze or precipitation and it's a whole new ball game. Aviation authorities in some countries flatly refuse to allow flight under VFR at night; you must file IFR. You can still fly VFR at night in the United States, but that doesn't mean it's necessarily any safer here. Save your VFR night flights for cloudless skies and full moons, when flying is almost a mystical experience.

Step 4. Call the intermediate and destination airports

Many pilots eliminate this step, or don't even think about it—there's no regulation that says you must do it or even hints that you should—but experienced pilots know it can save a lot of trouble. NOTAMs are not always current or even available to the briefer you consulted and there are often hazards that could affect the flight, but never make it into the "official" NOTAMs. A quick conversation with the local FBO might yield something like, "Watch out for the crane working over at the hotel on downwind to one-four, especially if you are coming in from the west." You never know.

Talking to someone local can also provide excellent information about the present weather conditions that might not be provided by automatic weather briefings and a briefer in a flight service station miles away from your destination. A report of VFR conditions at an airport might be useless if the mountain pass you plan to fly through to get there is totally obscured.

A telephone call is also a good way to confirm the information about the field that you find in airport information publications; the guides try to stay as up-to-date as possible, but publication lead-times are long and errata pages might be out-of-date.

Other things to check are that the field will be open when you arrive, that fuel will be available (a planned refueling stop is useless if the fuel truck is broken), that the runway lights will be working if you'll be landing after dark, and that the FBO will be open if you require any services. Uncontrolled airfields that are bustling with activity during the day often become dark and foreboding—and a long way from civilization—at night.

Hangar story

The private pilot planned to ferry a Cessna 152 Aerobat from Quakertown Airport back to Van Sant Airport where he had rented the airplane and left his car, a distance of about 20 miles. The wind was out of the northwest, about 20–25 knots, but because both airports had east-west runways, he figured the crosswind component would be minimal. Takeoff from Quakertown was easy, but the wind at Van Sant turned out to be more northerly and turbulent. After breaking off three approaches at 100 feet, the pilot decided it was too rough to land at Van Sant. A call on unicom from an instructor at Van Sant confirmed that no one was flying out of the airport that day because of the wind. After an uneventful landing at Quakertown, the pilot realized a quick phone call to Van Sant before takeoff would have saved him the flight and the temptation of making a difficult crosswind landing.

Step 5. Check the weight and balance and performance charts

I know it's a bore. I know you've made thousands of takeoffs and landings in the same airplane and nothing's ever happened. I know the manufacturer has figured in a cushion so it doesn't matter if you depart some pounds over maximum takeoff weight. (In truth, it does matter because it's a violation of the FARs, not to mention that you will have a longer takeoff run, a slower climb, a higher stall speed, increased fuel consumption, and reduced structural strength margins.) I know the performance charts are sometimes hard to interpret and doing weight and balance is unexciting. Maybe you prefer the excitement of trees rushing toward you as runway disappears and the airspeed indicator slowwwwwwly edges toward takeoff speed. That's the kind of excitement I'd rather avoid.

The importance of weight and balance is apparent in this quote from a Beechcraft pilot's operating manual: "The necessity for proper computation of the airplane's weight and balance cannot be overemphasized. In the basic design, it is planned that under normal loading the weight distribution of pilot, passengers, baggage, and fuel will balance the airplane for flight. Since these items are variables, it is possible to concentrate weight in such a way as to make the airplane unsafe for flight. The factors that must be considered in the weight and balance of the airplane are the installation of equipment after the airplane has been weighed, trapped or unusable fuel, engine oil, usable fuel, pilot and passenger weights, and baggage or cargo."

When the FAA and manufacturer determine the center of gravity limits for an airplane, they conduct considerable testing at the c.g. limits, but not too much beyond the published limits. If you exceed c.g. by only 1 inch outside the limits, you might inadvertently become a test pilot in that airplane. What's worse, you'll be conducting the test in a flight regime that the professional test pilots avoided because they quit their testing when the airplane began to exhibit minor stability and control problems.

Avoid too-far-forward and too-far-aft centers of gravity. A too-far-forward c.g. limits elevator control such that you might be unable to achieve enough nose-up attitude to take off. If you do manage to take off, you might not be able to hold a safe nose-up attitude for landing, particularly with flaps down and power at idle.

A too-far-aft c.g. is worse because you can end up with control instabilities that are beyond your ability to contain. You can get a nose-up pitch that can't be corrected with full-down elevator. If the airplane stalls, there might not be enough down-elevator authority to break the stall. Try to imagine pushing the yoke or stick full forward and your airplane continuing to stall. If that's not bad enough, the next obvious condition will be a spin, and with a too-far-aft c.g., there's no way you're going to get out of that.

As important as c.g. is, if there were an Aviation Ground School Subject Popularity Contest, "Figuring Weight and Balance" would probably finish dead last—and "How to Use Performance Charts" would be close by. No other two subjects consistently elicit so many moans, groans, and glazed-over eyeballs in the classroom. This applies to student pilots and professional pilots.

Weight and balance problems seem rather straightforward and simple, at the worst tedious. An electronic calculator eliminates most of the tediousness. A lot of the subject's bad reputation probably has to do with the fact that a small error done early in the calculations can cause an answer that is not only out in left field, but totally out of the ball park.

The best advice for anyone regarding weight and balance is to memorize the basic formula (moment = weight × arm), keep track of the labels that go along with the numbers (weight is pounds, arm is inches, weight times arm—and therefore moment—is in inch-pounds), and do the calculations carefully, writing down what you do as you go along. This will give you the right answer most of the time and help you find mistakes the rest of the time.

Weight and balance calculations are required for every commercial flight and the pilot in command must sign the form. If not computerized, forms covering every contingency are often filled out and copied ahead of time so that the form can be completed quickly prior to the flight. You can do something similar with your airplane or one you rent often.

Figure out the center of gravity for different circumstances. For example, hypothetically plan for one pilot with full fuel and full baggage compartment, or one pilot and three passengers with a half load of fuel and full baggage, or whatever. Keep these calculations in a small notebook with your other flight stuff so you can refer to them quickly.

One thing that is often overlooked by pilots is the fact that the center of gravity changes during the flight as fuel is burned off. This is particularly a problem in machines that have additional fuel tanks.

Last but not least, baggage compartments can be real culprits because they're often located in the forward or aft part of the fuselage. The center of gravity might be within limits with a full baggage compartment and a couple of passengers, but if the passengers hop out and the baggage is still on board, the c.g. could be out of limits. Just keep it in mind.

Hangar story

An Alaskan bush pilot flew his float plane to a lake to pick up some hunters that he had dropped off the week before and found they had two moose to take out. "I told you when I left that this airplane will only carry one moose!"

"The pilot who flew us in last year flew us out with two moose," one of the hunters responded.

Succumbing to the desire to preserve his image, the pilot agreed to take both moose. He packed everything, started the engine, taxied around to stir up the glassy water, and opened the throttle for takeoff. He held the airplane in ground effect for as long as possible to build airspeed before starting the climb. The airplane managed to clear the trees at the water's edge, but couldn't climb above the rise of the surrounding terrain. About a mile from the edge of the lake, the airplane mushed into the trees. Miraculously, no one was hurt.

Shaking his head, the pilot turned to his passengers and said, "I guess I just don't have what the other pilot did to get out of here."

"Oh, you did very well," one of the hunters said enthusiastically from the back seat. "The other pilot only made it a half a mile from the lake!" (Russ Lawton, AOPA)

The performance section of a pilot's operating handbook is a gold mine of information for the pilot who takes the time to do some digging (Fig. 1-2). The Beechcraft Sundowner manual, for example, has charts for takeoff speeds, climb speeds, stall speeds, landing speeds, wind components, takeoff distance—hard surface, takeoff distance—grass surface, normal climb, cruise performance, landing distance—hard surface, and landing distance—grass surface.

Many pilots of small airplanes can get away with not using performance charts very often because they operate from airports that have runways designed for much larger aircraft. With experience, they figure cruise performance using rules of thumb, which is acceptable under most conditions.

The time to pull out the performance charts to find out if your airplane will climb over the power lines at the end of the runway is not seconds after takeoff, when you

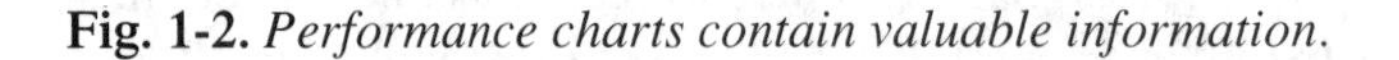

Fig. 1-2. *Performance charts contain valuable information.*

have to make a snap decision whether it's less risky to try to fly over the power lines or under them. Nor is it the time to look at the performance charts when you enter the traffic pattern at your destination airport with the engine sucking fumes and you wonder if the runway is long enough to accommodate your airplane. The time to do both is before takeoff. And it really does not take that much time. Remember, too, that FAR 91.103 does specify that you should "become familiar with all available information concerning the flight," including "the takeoff and landing distance data."

One word of caution. Pay attention to the conditions that pertain to the particular chart you're looking at. These can make all the difference in obtaining meaningful performance figures from the chart in question or figures that are completely wrong with respect to your present situation. It's very easy to open the manual, find a takeoff distance chart or table, for example, and start plugging in parameters. I've done it and I've seen other pilots do it. Check the conditions first! You don't want takeoff figures for the airplane at empty weight when you've loaded up to max gross.

Another word of caution. The performance charts are based on a new airplane with a new engine flown by an experienced company test pilot. The more time on an engine, the less power that it will develop. Compression decreases, plugs are fouled, and the carburetor isn't as efficient as it once was. In addition, it's likely that the parasite drag of your airplane is greater than on the airplane used to determine the performance charts. Dirt, rough paint, loose doors and fairings, and additional antennas contribute to increased drag. Remember, too, that a highly skilled test pilot did the flight tests that determine the performance charts, not an average pilot (if there is such).

How much of a cushion should you allow? How about 50 percent more than the charts? The Australian civil aviation authority actually adds 25–35 percent to performance data to aircraft built and certified in the United States, which provides an added margin of safety for the average pilot. It's your call.

Step 6. File a flight plan

If you plan to fly IFR, you must file a flight plan. But if you plan to fly VFR, it's optional. Unfortunately, many pilots regard VFR flight plans as an unnecessary infringement on their freedom. This is utter nonsense. When you file a VFR flight plan, you still choose where you want to go. If you change your mind, it's a simple matter to revise the flight plan in the air. Flight service stations are only too happy to help with a flight plan and provide simple flight following by recording any radio contacts; after all, that's part of the job.

Filling out a flight plan form is relatively easy, straightforward, and self-explanatory. The *Airman's Information Manual*, paragraph 5-4 explains it in detail if you have any questions. Figure 1-3 shows an example of a flight plan form.

Like checking the weather before departing, filing a VFR flight plan is an inexpensive insurance policy. If you are ever forced down short of your destination, it will be extremely comforting to know that someone will soon be looking for you, especially if you don't manage to get out a Mayday call.

U.S. DEPARTMENT OF TRANSPORTATION
FEDERAAL AVIATION ADMINISTRATION
FLIGHT PLAN

(FAA USE ONLY) ☐ PILOT BRIEFING ☐ VNR ☐ STOPOVER

TIME STARTED

SPECIALIST INITIALS

1. TYPE: VFR, IFR, DVFR

2. AIRCRAFT IDENTIFICATION

3. AIRCRAFT TYPE/ SPECIAL EQUIPMENT

4. TRUE AIRSPEED KTS

5. DEPARTURE POINT

6. DEPARTURE TIME: PROPOSED (Z), ACTUAL (Z)

7. CRUISING ALTITUDE

8. ROUTE OF FLIGHT

9. DESTINATION (Name of airport and city)

10. EST. TIME ENROUTE: HOURS, MINUTES

11. REMARKS

12. FUEL ON BOARD

13. ALTERNATE AIRPORT(S)

14. PILOT'S NAME, ADDRESS & TELEPHONE NUMBER & AIRCRAFT HOME BASE

15. NUMBER ABOARD

17. DESTINATION CONTACT/TELEPHONE (OPTIONAL)

16. COLOR OF AIRCRAFT

CIVIL AIRCRAFT PILOTS. FAR 91 requires you file an IFR flight plan to operate under instrument flight rules in controlled airspace. Failure to file could result in a civil penalty not to exceed $1,000 for each violation (Section 901 of the Federal Aviation Act of 1958, as amended). Filing of a VFR flight plan is recommended as a good operating practice. See also Part 99 for requirements concerning DVFR flight plans

FAA Form 7233-1 (8-82)

CLOSE VFR FLIGHT PLAN WITH ______________ FSS ON ARRIVAL

Fig. 1-3. *Sample flight plan form. A filed flight plan is the least expensive aviation insurance you can buy. It is free.*

You can easily file a flight plan by calling the universal toll-free number for flight service stations, 1-800-WX-BRIEF, and punching in "# 401" or "# 402" for Fast File. The 401 is for filing VFR and IFR flight plans (when the expected departure time is within 1 hour) and for closing flight plans; 402 is for filing flight plans when the expected departure time is greater than 1 hour from the time of filing. DUAT and other computer on-line services also have flight plan filing capability.

Step 7. Preflight the airplane

For some reason, as most pilots gain experience and flight time, they become more and more blasé with their preflight inspections. Apparently, the "it-won't-happen-to-me" attitude becomes stronger the longer one flies and nothing does happen. All pilots, private or professional, male or female, young or old, are susceptible to this attitude.

The macho, devil-may-care image most pilots emulate practically demands that one take a relaxed attitude toward preflights. In reality, the quick walkarounds of some pilots are not due so much to a high level of competency as to plain old laziness and complacency.

It's not hard to embarrass yourself with a less-than-sterling preflight. Once I was about to hit the starter of a rented Beechcraft Sundowner when the owner of the FBO suddenly appeared waving his arms for me to stop what I was doing. He made me get out of the airplane and pointed to the tow bar that was still fastened to the nose wheel. I had a reason for forgetting to remove the tow bar—brushing snow off the wings—but there really was no excuse. I was lucky he still let me rent his airplane and even luckier he'd caught me before I started the engine.

Recall that one of the first paragraphs in FAR Part 91 says "the pilot in command of an aircraft is directly responsible for, and is the final authority as to, the operation of that aircraft," which includes the airworthiness of the aircraft. It is impossible to verify if an aircraft is airworthy without inspecting it. Even if the aircraft has just been inspected and signed off by a mechanic, it's still the pilot's responsibility to ensure airworthiness before departing. Keep in mind that as good as your mechanic might be, everyone makes mistakes. Maintenance errors are inevitable because human error is inevitable.

What do you look for on a preflight inspection? Each airplane is different, so it's impossible to give you a specific checklist that would apply for every airplane. The one provided by the manufacturer is a good start, but often additional equipment requires you to customize it. You should add items. Shy away from any omissions because if the manufacturer put something in the checklist, there's a good reason for it being there.

The best, tried-and-true, used-by-thousands-of-pilots preflight inspection method is the aptly named walkaround (Fig. 1-4). You start at one spot on the fuselage and walk around the machine, checking things as you go. It doesn't matter if you start at the tail, or at the nose, or at the pilot's door. It doesn't matter if you walk around clockwise or counter-clockwise. It just matters that you do it the same way every time and you do it with genuine interest.

Fig. 1-4. *The "walkaround" is a time-honored way to preflight an aircraft, but it works best if it's not rushed.*

You do it as if, during the next hour or two, the condition of this machine will have a greater influence upon your life than nearly everything else in the world, because it will. Generally, you look for the unusual; an instructor should show you what is *usual*. It's also something you learn as you gain experience in your aircraft.

Checking crucial areas that are subsequently described form the basis for a thorough preflight. If any deficiency requires attention, you might consider minor corrective action on your own, but recall that you must follow the rules in FAR Part 43 regarding maintenance by the pilot or owner. Anything beyond Part 43 must be checked by an A&P.

Engine oil. If you fly your own airplane, you know when the oil was changed and when a quart was added. How much has been consumed? Color is usually a poor indicator of oil quality because after only a short time new oil becomes black, even in a relatively new engine, but smell can be a giveaway. If oil smells burned, it probably is burned. You can also check the oil breather pipe and the area on the fuselage behind it. If it's very oily, excessive oil is being forced out the breather, a sign that the crankcase is pressurized. This should be checked by a mechanic.

Landing gear and tires for condition. Tires should not have any cord showing and should be properly inflated. If you haven't checked tire pressure for awhile, do it—it's probably low. Look for leaks around the brake cylinders and scoring on the discs.

Fuel. You want to answer two questions: How much do you have? Is it contaminated? (Fig. 1-5) Ideally check the fuel visually through the filler hole; if not visually, at least read the gauges. If the airplane is not on level ground, FULL tanks might not be

Fig. 1-5. *Careful attention to the fuel that goes into your airplane is critical wherever you fly. In Barrow, Alaska, these pilots had to supply their own barrel, pump, and funnel. The fuel is being filtered through a clean chamois to remove water, dirt, and other impurities.*

full. It's surprising how much volume is lost by only a small amount of slope. You can do a simple experiment to illustrate this.

An empty 1-gallon plastic jug works well. Lay it on its side and pour as much water as you can into it. Turn it right side up and note how full it is. Depending on the jug, it will be about one-fourth to one-third full. Carefully lay it back on its side. Now tilt up the bottom slightly so that the water begins to flow out. You'll see that it only takes a small amount of tilt to lose a lot of water. Turn the jug right side up again and check how much water has drained out. Like the water jug on its side, fuel tanks in aircraft are usually long and flat, so you can see how a little slope can easily reduce the total volume of the tank.

Before refueling, check that the transfer valves between tanks are in the proper position. Electrically-operated valves probably will require that electrical power be switched on in order to be opened or closed. In many aircraft, fuel tanks are designed to feed by the force of gravity to other tanks. If both tanks are filled to capacity and the transfer valve is open, you can end up with fuel squirting out of the lower tank.

I was reminded of this the other day after I filled the wing tanks and half-filled the center tank in my Taylorcraft. On downwind for my second touch-and-go, fuel started spurting out from the center tank filler cap and all over the windshield. I quickly surmised that the wing tanks were feeding the center tank and I had forgotten to check the transfer valves. I closed the valves and the fuel promptly stopped squirting on the windshield. (The item was not on the written preflight checklist; it was inserted on the checklist after the flight.)

Drain some fuel from the sumps to check for water. Drain all the sumps—Are you sure you know how many your particular airplane has?—and use one of those clear, plastic cups so that you can actually look at the fuel instead of just polluting the ramp with it. (I often wonder how pilots get away with dumping raw fuel on the ground at airports all over the country without the Environmental Protection Agency catching on, but that's another concern.) Water will collect in the bottom of the cup. Keep taking samples until you don't get any water or other foreign matter.

If you're suspicious there's an excessive amount of water in the tanks, for example, because it's been raining a lot since the airplane last flew, rock the wings to slosh the fuel around, wait a few minutes for the water to settle, and take some more samples. Keep taking samples until you get two or three completely free of water. If you keep getting water, you should have a mechanic check the fuel system.

Know the color of the proper fuel (Table 1-2). If the color isn't right, don't fly!

Table 1-2. Fuel grade and color

Octane	Color
80	Red
100LL	Blue
100	Green
115	Purple (military use only)
jet	Clear or straw colored

Hangar story

A Bell Model 212 twin-engine helicopter had landed on an offshore platform to deliver passengers and pick up new ones to return to shore. The pilots took a fuel sample to check for the presence of water, supervised the refueling, and then went down to the galley for lunch. After eating, they returned to the helicopter, loaded the passengers and baggage, and started the engines.

Just before they were about to lift off from the platform, the first engine they had started flamed out for no obvious reason; seconds later the other engine flamed out. The pilots suspected fuel contamination and checked the aircraft's fuel drain. Water streamed onto the helideck.

It turned out that the platform's fuel system had somehow malfunctioned, causing sea water to be sucked up the overflow drain and into the main tank. Instead of being refueled with Jet A, the helicopter was "fueled" with salt water. The fuel sample taken prior to the refueling was representative of the Jet A in the hose, not the tank. The engines ran normally until the fuel in the fuel lines was used up. If the pilots had rushed through the checklist and managed to take off, they would have been forced to ditch. As a consequence, company procedure was changed to include a fuel sample taken after refueling, too.

Landlocked FBOs aren't in danger of sucking up sea water into the fuel storage systems, but it is possible for fuel contamination to occur. After any refueling, take fuel samples from each drain and don't be in a hurry to rush off into the wild blue yonder. Give the engine some time to run on the "new" fuel to make sure it's all right. It takes very little water to make a reciprocating engine quit and it won't start running again until it gets pure fuel. A sputtering engine on takeoff is not fun.

One final word on fuel: Contaminated fuel causes more engine stoppages than faulty magnetos, blocked air filters, and mechanical failures combined.

Propellers. Remember that propellers are airfoils. Smooth airfoils work better than ones that are nicked and scratched. Nicks should be filed out by a mechanic. Pilots aren't supposed to do this because improperly repaired nicks can turn into stress areas that can cause part of the blade to break off. Propeller blades have broken in flight because of damage that looked superficial. This is understandable when you realize that the propeller has more pressure exerted against it than any other component of the airplane.

Check that the blades of variable-pitch props are secured to the hub. If they're at all loose, check with a mechanic.

Windshield. I often forget this simple, but important item until I climb into the cockpit and the visibility changes from CAVU to "a mile in haze" because of the dirt on the windshield. You need a clean windshield no matter what the weather. Okay, if it's already raining, Mother Nature will take care of it for you, but in other conditions, at least start out with it clean. Windshield cleaning products are available, but I find that a clean cotton cloth and Pledge or similar furniture cleaning polish works great on plexiglass. It was recommended to me by an A&P mechanic who works for a commercial operator.

Wings, flaps, and tail. Check the security of nuts, bolts, cotter pins, and safety wire. If the linkages look dry, lubricate them according to the manufacturer's recommendations. Check for loose rivets on the skin; one sign of a loose rivet is a black streak on the skin behind it. Hairline cracks and stress lines should be checked by a mechanic, particularly if someone else flies the airplane. You never know what maneuvers the airplane has been put through. Of course, if you have been doing some wild stuff in the airplane yourself and you find some signs of stress, definitely have a mechanic look at them.

Circuit breakers. Check that all the circuit breakers are in. If some are popped, ask why. Don't just mindlessly push in a popped circuit breaker. It could have been pulled on purpose during maintenance or popped because of a temporary electrical surge, but it could be the indication of a failed component or system. Check the maintenance log for service on that system. If you can't determine a specific reason for the breaker to pop, make a note of which one it is. A popped breaker is normally not reason enough to cancel a flight, but if you really need the system for the flight you might want a mechanic to look at it. At the very least, be sure to check the breaker after that particular circuit is energized to ensure that it's in. If it pops again, you definitely should determine the cause.

Survival equipment. This will depend on your route of flight and might be as simple as a bottle of water and a first aid kit or as extensive as a life raft, antiexposure suit, and complete water survival kit. More information about what to carry is in appendix A.

Miscellaneous. Look for unfastened or missing fasteners and screws, open inspection doors, and excessive fuel and oil streaks. Remove all temporary covers and pins. Check the pitot tube, static ports, and fuel vents for obstructions. Spring and summer are times when insects and birds like to build nests inside protected areas in aircraft (Fig. 1-6).

Fig. 1-6. *Numerous bird droppings on the propeller and cowling of this airplane are a good clue that a feathered friend has found a home. Sure enough, a large nest was in the engine compartment.*

Make sure the fuel caps, oil caps, and baggage doors are secure. Check that some parts don't jiggle at all, others jiggle just a little, and others don't jiggle too much. The pilot's operating handbook for your particular airplane should tell how much jiggle to expect with different parts. Look for bends, cracks, dents, and defects. Look for foreign objects that shouldn't be there: screwdrivers, wrenches, pieces of safety wire, and the like. A Sundowner pilot, for example, found a plastic bottle wedged between the engine and firewall to catch oil from the oil filter. It had been left there by a mechanic after an oil change and contained about half a quart of oil.

Hangar story

This miscellaneous category is not insignificant. You've probably heard the old tale about the nail and the battle: for want of the nail a shoe was lost, for want of the shoe a horse was lost, for want of the horse a rider was lost, and for want of the rider the battle was lost. Things like that happen in aviation. An Embraer 120 Brasilia was lost because of a few missing screws.

The commuter airplane was scheduled to have the deice boots replaced on the leading edge of the horizontal stabilizer, which, as part of its T-tail, is about 20 feet above the ground. To replace the boots, the leading edge of the horizontal stabilizer must be removed from the airplane. Two mechanics started on the right stabilizer, removing the lines of screws on the lower side of the leading edge. One of them climbed on top of the stabilizer and removed the screws from the upper leading edge of both the right and the left stabilizers.

The job was not finished when the shift change took place at midnight and the mechanics on the new shift did not know the screws holding the upper leading edge on the left side had been removed. The right boot was overhauled, but lack of time caused overhaul of the left boot to be deferred until later. No one noticed the missing screws on the upper surface of the left stabilizer, or thought to look, because all the screws were in place on the lower surface of the left stabilizer.

The aircraft was on its second flight of the day when the leading edge on the left stabilizer failed and separated from the aircraft due to flight loads. The subsequent loss of downward lift on the tail caused an abrupt nose-down attitude of the airplane of about 4–5 Gs. The airplane disintegrated in flight while passing through 12,000 feet at a speed of 260 knots. The left wing broke (although it remained attached to the fuselage), the left engine separated, and then the horizontal stabilizer with part of the vertical fin separated, too. From the time of the abrupt nose-down tuck until the flight recorder stopped recording, 30 seconds elapsed. All 14 on board were killed.

So don't ignore those small screws, fasteners, and other miscellaneous items.

Your first flight instructor should have shown you how to make a thorough preflight inspection. If not, and you still wonder how to do one properly, find another instructor or an experienced pilot you trust, and ask to be shown. After you have some time in your logbook, it's a good idea to ask an instructor to show you a complete preflight again. You'll be surprised how much more you understand. Things that appeared to be meaningless actions will acquire new significance—and now you'll remember them much better. When you check out in a different airplane, be sure to get a thorough preflight from a pilot experienced in the machine.

Then it's up to you to be a professional about your preflight inspections every time, whether you fly for hire or for fun.

Step 8. Brief your passengers

It's easy to forget to brief your passengers, especially if you fly with the same people often—in which case it might not be necessary to brief them every time—but anytime you fly with someone new, it's always important to give them a thorough briefing about what they can expect during the flight. Even frequent travelers like to know what's going to happen and what you expect of them. Many of your passengers will no doubt be experienced airline passengers and be able to recite the litany about seat belts and oxygen masks, but they might not be familiar with smaller airplanes, even the larger turboprop twins; therefore, it's always a good idea to explain the particular safety features of your airplane, give your passengers an idea what to expect, and tell them anything you'd like them to do or not to do.

Remember, too, that first-time passengers are usually nervous, even though they might appear very enthusiastic and disavow any anxiety. They will appreciate a good briefing; however, don't neglect the passenger briefing just because your passengers are frequent light airplane fliers or pilots. It's always better to repeat the obvious than to assume, erroneously, that someone already knows what you think they know. When you fly with other pilots, be sure to clarify who is the pilot in command and what is expected of the second pilot.

You might want to customize your own briefing and print it on a card for easy reference in the cockpit. The following items are appropriate for most passengers in most small airplanes:

- Outside the aircraft. Tell your passengers that smoking is not allowed on the ramp and they should stay well clear of the propellers, even when the props aren't turning. When boarding the airplane, passengers should step only where indicated and use the appropriate handholds.

- Seat belts. Demonstrate how to fasten and adjust the seat belts and shoulder harnesses, if installed. Instruct passengers to keep belts fastened at all times.

- Doors and windows. Demonstrate how to open the doors and windows, but instruct passengers only to open a door in the event of a crash landing. Otherwise, the pilot will always open the doors for them. Show passengers how the window vents work and when they may be opened.

- Cockpit courtesy. Tell your passengers there are times when it will be necessary for you to concentrate 100 percent on flying and should not be distracted or disturbed, except for something crucial. There should be no extraneous conversations during takeoff, climbout, descent, and landing. Tell passengers that anytime they see you talking on the radio or hear a radio transmission, they should be quiet. Also, if the flight will be transiting high-traffic areas, such as Category B or C airspace, conversations should be limited to necessary information only, so that radio calls are not missed.

- See and avoid. Explain the importance of looking for other air traffic and advising you of it. This can be turned into a game with children (and some adults) by giving a prize to the person who sees the most airplanes first.
- Controls and instruments. Instruct passengers not to touch any controls or instruments unless you give them permission to. You should also tell them that some instruments might blink or make a noise periodically and this is normal. For example, the transponder blinks when it is interrogated by ground radar and the stall warning horn might sound on landing.
- Weather. Be honest with your passengers about the weather. If it's going to be turbulent, let them know, and reassure them that "bumps" don't hurt the airplane—at least not the bumps that you expect during the flight. (If you expect airplane-breaking bumps, you shouldn't be going.) Be aware that what most pilots call "a little turbulence," many passengers would classify as "a lot of turbulence." If you're going to be flying IFR and in the clouds, let your passengers know and briefly explain to them how you are able to fly right side up and navigate when you can't see outside the windows.
- Airsickness. Let your passengers know that it's no embarrassment to become airsick; it can happen to anyone, even experienced pilots sometimes. Advise them to tell you if they start to feel queasy and show them where the airsickness bags are located. As with any motion sickness, remind them that it is better to look outside than inside and that reading might trigger airsickness.
- Heating and ventilation. Show your passengers where the vents are positioned and how to adjust them.
- Pressure changes. Explain how climbing will cause ears to "pop" and that descending might cause ear pain. Have chewing gum available and explain how to release the pressure building up in your ear channels by holding your nose and mouth closed and blowing out gently through the ears. Note: With a cold or ear infection, the increase in pressure can become very painful and difficult to alleviate; anyone so afflicted should not be flying, especially as pilot in command.
- On landing. Remind passengers to keep their belts fastened until you've stopped the airplane and shut down the engine. If they will be disembarking while the engine is still running, remind them to avoid the propeller.

Step 9. Organize the cockpit.

When your passengers are settled, you can get yourself organized. There are probably as many organizational methods as there are pilots and you can even buy "cockpit organizers" from catalogues that specialize in pilot supplies. The size of your cockpit and whether or not you have a passenger sitting next to you are two factors to consider. Many pilots like to use a kneeboard that straps to a leg.

As a minimum, you want to have handy the aeronautical charts you'll be using, an E6B computer or electronic equivalent, a couple of pens or pencils, a clipboard or kneeboard to write on, your flight planning log, an *Airport/Facility Directory*, *Flight-Guide*, or *AOPA's Aviation USA*, sunglasses, a hat with a visor, and a flashlight. Even for daylight-only operations, a flashlight is useful for looking in dark places in your airplane and a precautionary landing could put you in an unlit field after dark. For an IFR flight, you should obviously have the instrument charts and all the approach plates for the area. Add other items you find you need.

Stash items you need in flight where they are easy to reach. Put nonflight items out of the way: fuel-testing cups, tiedown ropes, and the like. Develop a system that works well for you and then stick to it. If you always keep your flight log on the left and your E6B in the small compartment on your right, for example, you'll be able to find them quickly when you need them.

Step 10. Start the engine and go

This step is covered in detail in the next chapter.

COLD WEATHER CONSIDERATIONS

Like most machines, aircraft work better when they're warm. Although colder temperatures offer some distinct advantages, such as low density altitudes, they also present special problems.

Cold weather is relative. A winter in Florida and Arizona is a lot different from a winter in Vermont and Montana. Summers in Alaska and Canada often seem like winters in the "Lower 48." And in many parts of the United States, the temperature can dip below freezing 10 months out of the year.

The best advice you can get about winter operations is from operators and mechanics permanently located in the area. The information below will apply to some areas and not others. If you'll be spending an extended time in a cold-weather area, seek the advice of the locals.

The best place to park an aircraft in cold weather is in a warm hangar (Fig. 1-7). Unfortunately, this isn't always possible. An alternative, if you really don't need to fly, is to winterize your airplane and not fly it when the weather is cold. I asked one Alaskan pilot what special things he did with his airplane for winter operations and he said, "I remove the battery, drain the oil, tie it down good and tight, and wait for the spring thaw."

I am going to assume you want to fly in winter, so here are some tips.

Engine preparation

Certain things can be done to prepare your aircraft for cold weather operations. Changing to winter-grade oils and greases is important to ensure the proper viscosity at lower temperatures. Insulate oil lines, oil pressure lines, and tanks with fireproof insulation to help prevent the oil from congealing. Baffles, cowling covers, and oil cooler covers are recommended by some manufacturers.

Fig. 1-7. *The best place to warm up an aircraft prior to a flight in cold weather is in a hangar.*

Congealed oil can also cause problems with oil-pressure controlled propellers. The installation of a recirculating oil system for the propeller and feathering system is helpful in extremely cold climates. Caution should be taken when intentionally feathering the prop during training to ensure that the propeller is unfeathered before the oil in the system becomes congealed.

Particular attention should be paid to the oil breather in reciprocating engines. When vapors in the crankcase cool, they might condense in the breather line and subsequently freeze, closing the breather. Frozen breather lines can create numerous problems.

Flexible rubber hoses, hydraulic fittings, and seals become brittle at cold temperatures, often breaking and causing leaks. After replacing unserviceable components, clamps and fittings should be torqued to the manufacturer's cold-weather specifications.

Battery

If the aircraft is to be left outside for a long period, the best thing to do with the battery is to remove it and keep it in a warm place. Keeping the battery fully charged with periodic recharging might be necessary because the starter will pull more power to start a cold engine and a cold battery will have less power to give. An external power cart might be the only way to start the engine in extremely cold temperatures.

Cables

Control cables are subject to contraction and expansion due to temperature changes. These should be checked often and adjusted to the proper specifications.

Wheel pants

Because of the possibility of mud or slush freezing wheels and brakes, it might be necessary to remove the wheel pants from fixed-gear aircraft. Wheel wells of aircraft with retractable landing gear should be inspected and cleared of slush and mud after each flight.

Preflight tips

The urge to hurry the preflight inspection when it's cold outside is natural, particularly when the aircraft is outside and adverse weather conditions exist. This is the very time to execute the most thorough preflight. If you dress for the conditions, you'll be less inclined to hurry. With a thought to survival, if you're not dressed warm enough to do a 10-minute preflight, how are you going to stay warm for hours or days after an unplanned landing?

All frost, ice and snow must be removed from the airfoils and control surfaces, and from around the static system sensing ports (Fig. 1-8). Preheating in a warm hangar is most effective technique for ice removal, particularly if the aircraft is left inside long enough to dry. If you normally park your airplane outside, try to get it into a hangar for

Fig. 1-8. *All snow and ice must be removed from the wings and ailerons before this airplane will be fit to fly.*

a few hours before you fly. If that's not possible, alcohol or one of the commercial ice removal compounds can be used to remove ice and snow. Be sure water doesn't run into control surface hinges and crevices only to freeze later.

If your aircraft is parked outside in blowing snow, pay special attention to openings where snow can enter, freeze solid, and obstruct movement of control cables and the like. These openings should be free of snow and ice before the flight. A few days after the Blizzard of 1993, I was surprised to find a large block of snow inside the tail of my Taylorcraft (Fig. 1-9). Snow had also blown into the cabin. (I will admit that the doors don't seal well.) Be sure to check pitot tubes, static ports, wheel wells, heater intakes, carburetor intakes, tailwheel area, and fuel vents.

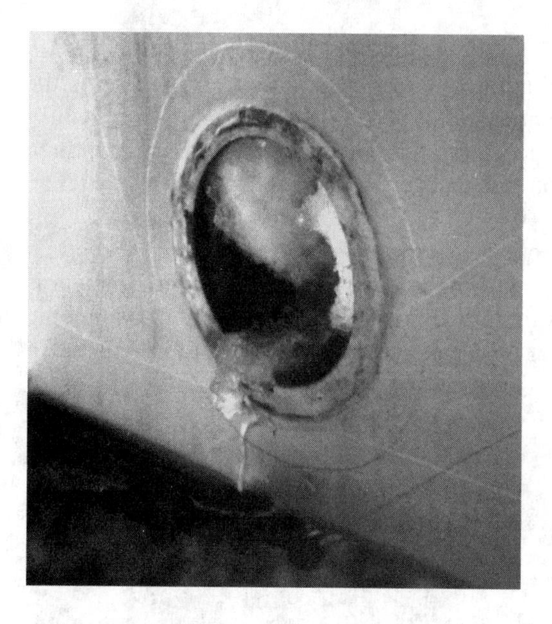

Fig. 1-9. *After a blizzard that included much blowing snow, I found snow had blown* inside *the aft fuselage of my Taylorcraft. If I had not noticed water dripping from the tail, I would not have removed the inspection cover to investigate. Obviously, snow compacted around the control cables could have caused a serious problem in flight.*

If nothing comes out of the drains when you drain the fuel, there's a good chance the water in the drains has frozen. To be safe, get the airplane into a hangar so the fuel temperature warms up above freezing.

Extra care should be taken during changes in temperature, particularly when it nears the freezing level. Ice in the tanks might turn to water as the temperature rises, and pass through the filter into the carburetor or fuel controller causing the engine to stop. During freeze-up in the fall, water can freeze in lines and filters, causing engine stoppage and fuel leaks.

Cold-soaked fuel

One often overlooked problem with cold weather operations is cold-soaked fuel. Any object becomes "cold-soaked" if exposed to a cold temperature long enough to drop to that temperature. Obviously, the fuel and everything else in an airplane that is parked outside will eventually attain the outside air temperature (Fig. 1-10). Not quite so obvious is the fact that, given enough time, the fuel in an airplane's wings can also cold-soak while in flight. With respect to engine operation, this is not a big problem as long as the fuel is not so cold that it no longer vaporizes; however, fuselage icing can become a big problem if the aircraft encounters above-freezing temperatures and precipitation.

Fig. 1-10. *In addition to removing the snow and ice from the wings and fuselage of this Cessna 172, the pilot should also be concerned with problems of cold-soaked fuel.*

The icing problem can occur on the ground, as well. Assume a cold-soaked airplane is about to depart on a flight. The ambient temperature is slightly above freezing and it's either raining or snowing. Normally, the skin temperature of the airplane will be close to the air temperature and there will be no danger of ice forming; however, the airplane's fuel, which changes temperature much more slowly than air or a thin layer of metal, is still below freezing. If there is still enough fuel in the tanks to maintain contact with the skin, the skin temperature will be brought below freezing. (If you refuel from a cold-soaked fuel truck, the same thing will happen.)

Any precipitation in this situation will freeze and stick to the wings. This kind of ice is just as bad, perhaps worse, than in-flight icing and should be removed before takeoff. The takeoff and climb crashes of an Air Ontario Fokker F-28 in March 1989 and a USAir F-28 in March 1992 were partly caused by cold-soaked fuel that resulted in severe icing on the wings.

Even slushy snow should be suspect. During the takeoff roll, the additional cooling of the air flow over the wings could be enough to cause almost-frozen water to lose a few calories and crystallize into ice. Don't take any chances: Clear the airplane of snow, slush, and ice before takeoff and if you suspect your airplane's carrying cold-soaked fuel, don't fly into wet snow or rain.

Other cold problems

Low temperatures might cause a change in the viscosity of engine oils, batteries might lose a high percentage of their effectiveness, and instruments might stick (Fig. 1-11). Preheating of the engine and cabin before starting is desirable, if not downright necessary in low temperatures. The best way to preheat is in a heated hangar. If you use a portable heater, do not leave the aircraft unattended and always have a fire extinguisher nearby. Be careful not to direct the heat on flammable items, like upholstery. Preheating takes a long time to be effective. Obviously, the colder the outside temperature, the longer preheating will take.

Fig. 1-11. *Even if an airplane is protected from the elements by a shelter, preheating of its engine when the temperature is below freezing is beneficial.*

Hangar story

A 55-year-old pilot with a private pilot license and more than 4,000 hours drove to South Lake Tahoe Airport, California, arriving shortly after 7 a.m. The sky was clear, but the OAT was 22°F. He found his Cessna 401 twin-engine airplane covered with a thick coat of frost, up to one-half inch in places.

Witnesses said afterward that the pilot did a very minimal preflight, only scraping the ice off the windshields. He and his two passengers then boarded the airplane and within minutes, the Cessna's 300-hp Continental engines started up. The pilot taxied directly from the tie-down spot to Runway 36, never stopping to warm up the engines or do the normal run-up checks.

Shortly after takeoff, the left engine lost power. The pilot decided to return to the airport, and turned left, toward the "dead engine." This might have contributed to his subsequent inability to cope with the asymmetric thrust from the right engine. Actually, it would have been better if he had waited to turn until the airplane had climbed to a higher altitude and he had sorted out the emergency. He neglected to feather the windmilling left propeller, which would have reduced drag on the left side and improve directional control, but he did put the gear down. This is a mixed blessing with a single-engine failure in a twin; lowering the gear improves directional stability, but also increases drag.

The combination of asymmetric thrust from the right engine, drag from the windmilling left propeller and lowered landing gear, and deterioration of the aerodynamic effectiveness of the wings because of the layer of frost quickly caused the airplane to stall, roll inverted, and crash. All three persons on board died.

The NTSB could find no mechanical problems that would have caused a power loss in the left engine. If the pilot had let it warm up prior to takeoff, it probably would not have failed. If he had cleaned the frost off the wings and followed the proper procedure for an engine failure after takeoff, he probably would have landed safely, even if the engine had failed.

BE PREPARED

Most new pilots religiously perform all of these preflighting steps. Then, as the total of hours in the logbook builds up into the hundreds, and eventually the thousands, they tend to become somewhat more casual about their preflight preparations. Some experienced private pilots see nothing wrong with this because they feel that the more they fly, the less likely they are to be confronted by situations that they cannot easily handle based on past performance. On the other hand, professional pilots take a different point of view and tend to take things much less casually.

Most professional pilots stack the cards in their favor as much as possible: double-checking the weather, adding extra fuel, and using checklists. As the old saying goes, "The superior pilot uses superior judgment to avoid having to use superior skill."

The military, always looking for easy ways to remind pilots how to do the right thing, came up with the 6 Ps (Fig. 1-12): Proper Preflight Planning Prevents Poor Performance. It might sound a little corny, but the message is valid.

Remember: The longer the flight, the greater the chance that something unexpected will occur. The better prepared you are before takeoff, the easier it will be to meet any adverse condition.

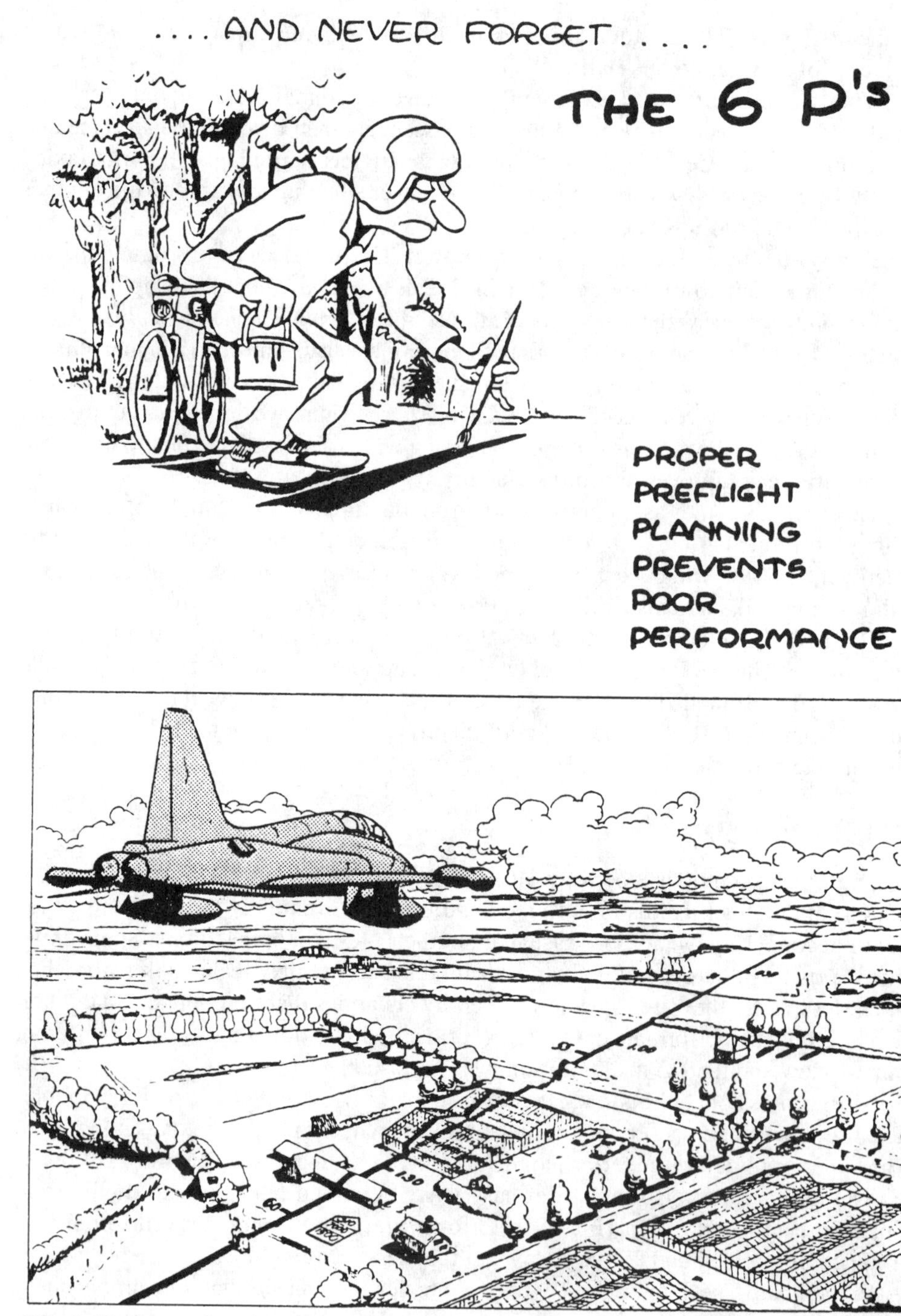

Fig. 1-12. *Remember the six Ps: Proper Preflight Planning Prevents Poor Performance.*

2
Engine start, takeoff, and departure

"Twenty tries at takeoff had won me only the understanding of the power of a snowflake, multiplied a thousand billion times. The heavy wet stuff turned to thick soup blurring underwheel, blasting violent hard foundations against the struts and wings of my borrowed Luscombe. Full power would drag us up to thirty-nine miles per hour at the fastest, and we need forty-five minimum to take off."

Richard Bach, ***A Gift of Wings***

USING YOUR BEST JUDGMENT, YOU'VE MADE THE GO-NO-GO DECISION—IT'S a go. Now it's time to see what that entails. Following the checklist carefully, you've come to engine start. In any powered aircraft, the engine is one of the most important elements, so let's look at it closely.

ENGINE SMARTS

You might have heard the radio commercial for a popular brand of engine oil treatment that says the worst thing you can do to your car's engine is to start it. Using the treatment is supposed to help. I don't know if it's any good, but I do know that what they say about starting the engine is true. You could preserve your engine forever if you never started it, but that defeats the purpose of having a car or an airplane in the first place. Because you do have to start the engine, what can you do to keep damage to a minimum?

Oil during engine start

When an engine has been sitting quiet for awhile, gravity exerts its relentless force on the oil and pulls it to the pan at the bottom of the crankcase. The pistons, cylinders, camshaft, valve tappets, crankshaft bearings, and everything else inside the engine will be virtually dry of oil if the engine sits long enough. With the oil in the sump, it takes time for the oil pump to start delivering oil to the system. As soon as the starter turns the engine over, all these parts will be rubbing against each other without the benefit of a lubricant between them.

Making the situation worse is the outside air temperature. When organic oil is cold, it thickens up and does not flow as well as when it is hot. If the oil is too thick, it will simply be impossible for the oil pump to move it. This is called *cavitation*. Oil passages become plugged, spray ports dribble oil instead of squirting it, and splash-lubricated components don't get splashed.

Synthetic oils, which are formulated to maintain viscosity at lower temperatures, don't thicken up as much as organic-based oils. Multigrade organic oils, such as AeroShell 15W-50, work better at a wider range of temperatures than single-weight organic oils. For example, a straight 50-weight oil, which is a summer-grade oil, just about refuses to flow when its temperature gets down toward freezing.

Therefore, before you do anything else, make sure you have the right oil for the time of year. You do this by following the engine manufacturer's recommendations.

Because oil is also used to cool the engine by reducing friction and radiating away heat as the oil circulates, you will end up with a very hot engine if the oil doesn't circulate. When this happens, the odd thing will be that the oil temperature gauge will still indicate a very low temperature because the temperature bulb is usually located in the oil sump and the oil in the sump will remain cold even though the rest of the engine is heating up quickly. If an engine runs long enough without oil, it will eventually seize. When it's very cold outside, the best thing you can do for the engine is warm it up before start. The second best thing is not to start it at all.

Hand-propping cautions

If the engine is only moderately cold, you can start the oil circulating by turning the propeller through by hand a few times. Because the oil pump is driven by a shaft connected directly to the crankshaft, when you turn the prop, you turn the crankshaft and the oil pump. This will move some oil throughout the engine before you start it. If the oil is so cold that it flows like molasses in January (to use an old, but appropriate, cliché), it probably won't help much to turn the engine over by hand. In this situation, warming the engine prior to starting it is your only recourse.

Whenever you turn an airplane's propeller by hand, treat it as if it will spin on its own at any time. Make sure your feet are on solid ground, keep your body well clear of the arc of the propeller, and quickly move your hands away after pulling the prop (Fig. 2-1). A pilot knowledgeable with the operation of the airplane should be at the controls.

The magneto switch, which is often incorporated with the ignition/starter switch, should be OFF; however, even with the switch OFF, a loose or broken ground wire could make it possible for a magneto to energize and send a zap of electricity to the spark plugs. If this happens, the engine could kick over a few revolutions or even start;

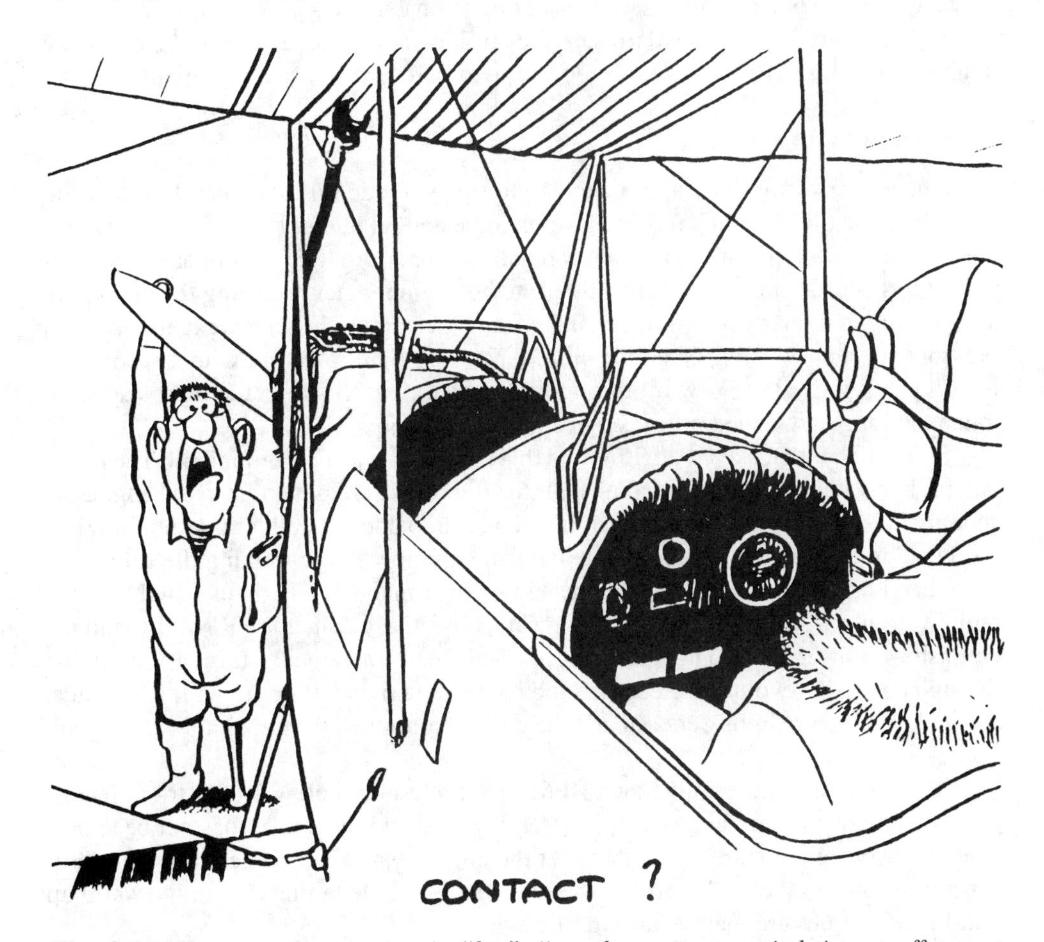

Fig. 2-1. *Always assume a prop is "hot" (i.e., the magneto switch is* not *off or* not *grounded) when turning the propeller by hand. Keep your body parts away from the propeller's arc and be ready to jump out of the way quickly if the engine kicks over.*

be sure to stay well clear. It doesn't take much to cause the engine to start. In one incident, a pilot had barely moved the prop six inches when the engine roared to life.

The throttle should be closed (fully pulled back away from the panel), the mixture control should be full lean or idle cut-off, and the main fuel switch or valve should be OFF or CLOSED. The parking brake should be on. Finally, the airplane should be secured to the ground; tied-down with the wheels chocked is the most secure, although tied-down is usually sufficient. Wheel chocks are better than nothing, but airplanes have been known to hop over chocks after being hand-propped.

If you can find someone familiar with hand-propping, it's usually best if you sit in your airplane at the controls because you probably know better the ins and outs of that airplane. But if no one else wants to turn the prop for you and you must do it yourself, be sure to give the pilot in the cockpit a briefing on where that airplane's magneto switch is, how the brakes work, and all other relevant operational aspects of that airplane.

If you are by yourself, be darn sure the airplane is well secured to the ground, the magneto switch is OFF, and the throttle is closed before you turn the propeller.

Hangar story

One winter many years ago, I decided to try synthetic oil in my car. In my desire to drain all the old, organic oil from the crankcase and oil pump, I let the engine run for a few minutes with the drain plug out. The oil low-pressure light illuminated, but I expected this because there was little oil left in the engine. After replacing the drain plug and pouring in 4 quarts of synthetic oil, I started the engine. From my experience with previous oil changes, I expected the oil low-pressure light to go out within seconds as the oil pump moved the new oil through the system, but this time the light stayed on much too long. I switched off the engine and thought about it.

Because the light stayed on, I figured that the pump wasn't pumping oil. The pump had probably cavitated because almost all the old oil was removed from the engine by running it. My dilemma was that a cavitated oil pump doesn't deliver oil to the engine, but I couldn't figure out how to "uncavitate" the pump without running the engine.

Then I hit on the idea of turning the engine over manually. Perhaps I could get the pump working without running the engine at a potentially damaging RPM. I found the crankshaft bolt at the front of the engine and, using a wrench, turned the engine through about 40 revolutions. I started the engine again and after only a few seconds, the oil low-pressure light went out and the engine continued to run without a problem.

After you start the engine, keep RPM low, preferably below 1,000, for at least a good minute or two and longer if it's extremely cold and the engine has not been preheated. If you're not sure how long to let the engine warm up, be generous and let it run a little longer at idle. The cost of the fuel you use while letting the engine warm up will be saved in future maintenance to the engine.

Your communication and navigation radios probably also would appreciate some warm-up time, so you might as well turn them on as the engine is idling. The gyros inside the heading indicator, attitude indicator, and turn coordinator can take up to 3 minutes to become fully functioning at standard temperatures, longer if it's really cold. Some manufacturers recommend that an airplane not even be taxied until the gyros are fully up to speed.

TAXIING

With the preflight planning accomplished, a thorough preflight inspection of the aircraft complete, and the engine or engines started and warmed up, it's almost time to take off. The engine run-up and magneto check need to be accomplished, and these should be done without hazard to people, buildings, or other aircraft. You might be able to do them on the parking ramp if you're out in the middle of nowhere—which is not uncommon for many airports—but in many cases you'll need to taxi to a run-up area. Taxiing an airplane is an operation that is more hazardous than most people realize (Fig. 2-2).

Fig. 2-2. *Taxiing an aircraft requires a pilot's full attention. Be extra vigilant when people and other aircraft are around, such as at a busy fly-in.*

Wind, surface conditions, and other aircraft are your biggest concerns. Sometimes all three seem to conspire to make your life miserable. Icy pavement, a strong crosswind, and airplanes parked tightly on the ramp can make getting to the runway a very hazardous part of the flight.

A single pilot is at a disadvantage. With two crewmembers in the cockpit, one pilot takes care of aircraft control while the other does all the other stuff: reading the checklist, talking on the radio, copying clearances, and the like. When only one pilot is doing the same tasks as two, she's got to take more time or something is going to be missed. Whenever the aircraft is in motion, in the air or on the ground, the most important task is always aircraft control.

Like driving a car, as soon as you put your aircraft in motion, you must pay attention to what you're doing. Unfortunately, unlike driving, there are extra distractions, such as radio calls and checklists, that can easily draw your attention away from the primary task at hand—taxiing.

Human engineering

A few human engineering factors related to taxiing tend to work against us. Most of us drive cars before we fly airplanes and we have learned to steer with our hands, accelerate with our right foot, and stop with our left. It's only natural that we might transfer these habits to the cockpit of airplanes, particularly when taxiing.

As any student pilot knows, learning to steer a vehicle with two feet and operate a throttle (accelerator) with a hand is a totally new and perplexing experience. The main problem is we've all learned to do one thing with our hands and feet while driving a car and now we have to do totally different things.

Because we don't stop driving cars after becoming pilots, we have to learn to switch from one to the other. For most people, this becomes easier as they gain more flight time and professional pilots who fly frequently usually make the adaptation easily and without thought because they do it often. But for less-frequent flyers, the problem can be more acute.

The brain does funny things to us sometimes, particularly the unconscious part of the brain. You could go for years steering your airplane with your feet and never once inadvertently turn the yoke when you want to turn the airplane. Then it happens: You're in a rush to take off, you start the engine, get clearance to taxi, and as you're rolling and finishing up the checklist, maybe entering a route in the loran or GPS, and ground control calls to tell you to switch over to tower frequency.

Your hands are flying all over the cockpit and your mind is on frequencies and lat/long coordinates and you need to make a right turn onto the taxiway ahead, so you instinctively push the right pedal to start the turn. Halfway through the turn you look at the yoke and realize you've also rotated it to the right—as if you were driving a car. "Gads! How'd I ever do that? I've got 5,000 hours and I made a mistake like this?" No damage is done and nobody saw you do it, but it makes you wonder.

Your unconscious mind acting on its own made you do it because the conscious part of your brain was too occupied with everything else that was happening. Because you're able to taxi almost without thinking about it, you can safely allow yourself to do it. When you bury those thoughts too deep, the unconscious will be making decisions on its own, and the unconscious draws on experience. No matter how many hours most of us spend flying, we've spent many more hours behind the wheel of a car. That means the preponderant memory we have about vehicle operation is driving an automobile.

When you want a car to turn right, you turn the steering wheel right. When taxiing an airplane, your hands are on the wheel-like yoke, your feet are on pedals, and you're rolling on the ground. With these sensory inputs, it's easy for your unconscious to assume you're in a car and quite naturally tells your hands to rotate the wheel clockwise. Normally, your conscious mind is able to correct this erroneous assumption before your hands react, but when your conscious mind is involved with numerous other things, it just can't catch the error in time.

So stay focused while taxiing, especially when conditions are less than ideal. Take your time and pay attention to the task at hand. Do the checklists and systems checks while stopped, not while moving.

Wind correction

Proper manipulation of the controls is important whenever wind is a factor, whether it's natural or man-made. The rule of thumb for tricycle-gear airplanes is to turn *into* a quartering headwind (upwind aileron up) and hold the elevator neutral (Fig. 2-3). With a quartering tailwind, turn *away* from the wind (upwind aileron down) and hold the elevator down (stick forward).

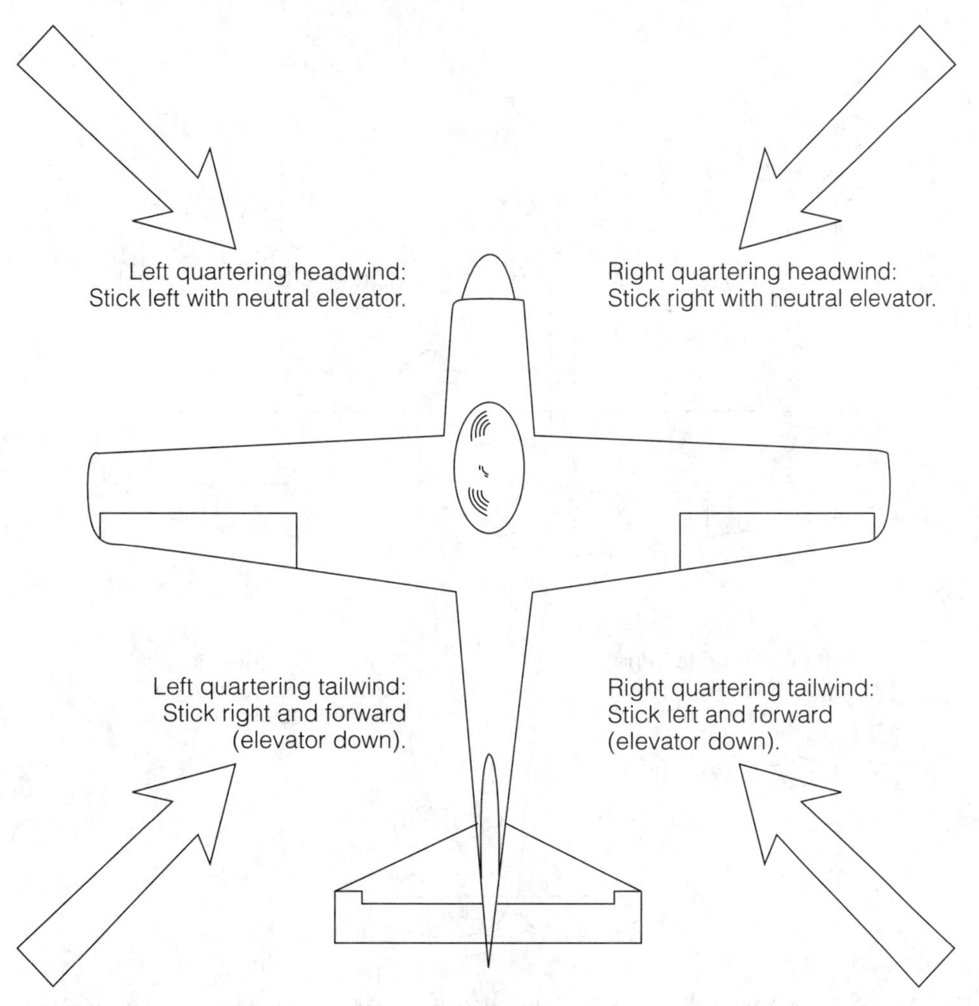

Fig. 2-3. *Rules of thumb for compensating for wind when taxiing a tricycle-gear (nose-wheel) airplane. Turn* into *a quartering headwind and hold the elevator neutral. Turn* away *from a quartering tailwind and hold the elevator down.*

The technique is slightly different with a tailwheel airplane (Fig. 2-4). With a quartering headwind, the stick should be turned into the wind (upwind aileron up), but the elevator should be held up (stick back) in order to keep the tail on the ground. With a quartering tailwind, the procedure is the same as with a tricycle-gear airplane: turn *away* from the wind (upwind aileron down) and hold the elevator down (stick forward). Figure 2-5 shows the effect of a tailwind on a wing and aileron or a horizontal stabilizer and elevator.

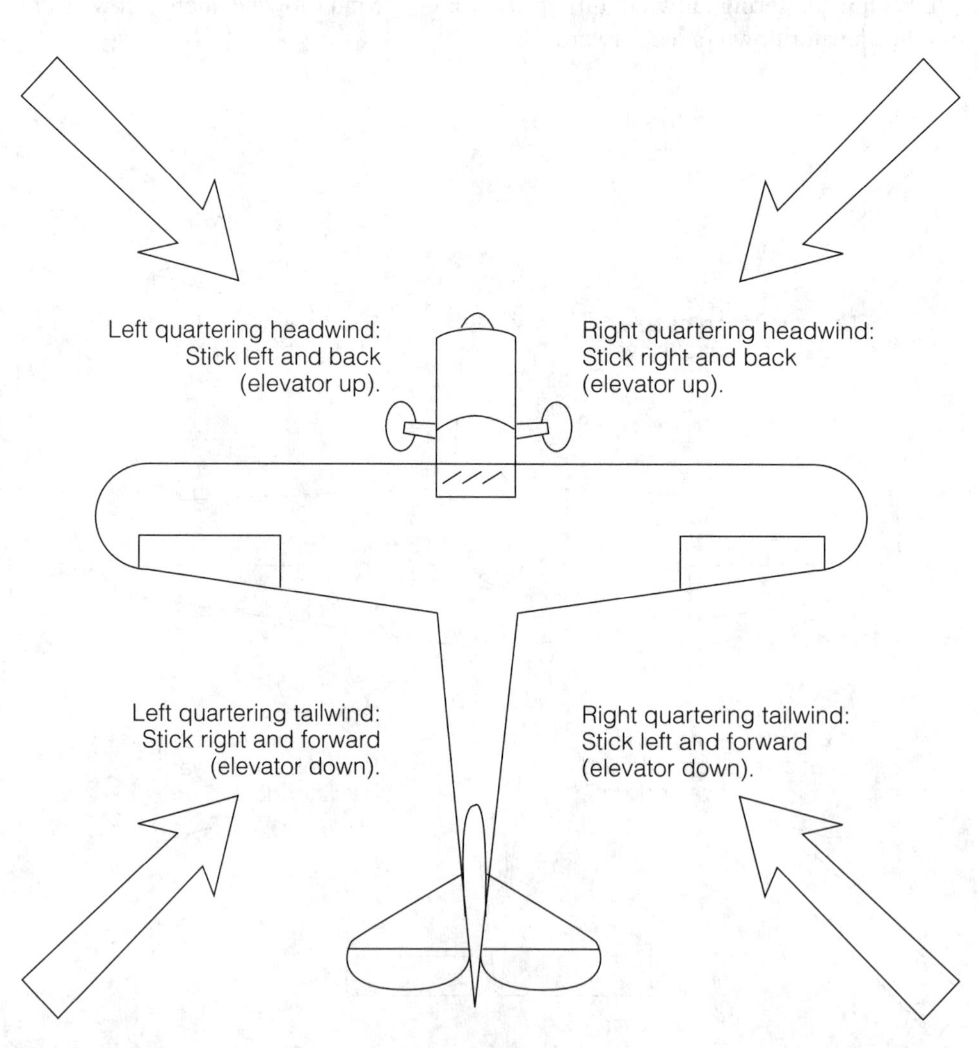

Fig. 2-4. *Rules of thumb for compensating for wind when taxiing a tailwheel airplane. Turn* into *a quartering headwind and hold the elevator up. Turn* away *from a quartering tailwind and hold the elevator down.*

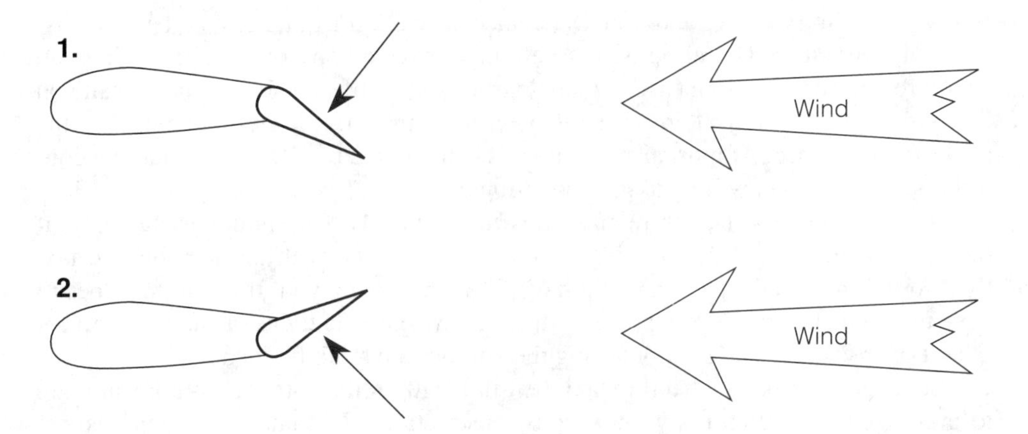

Fig. 2-5. *Lifting effect of a tailwind on an airfoil with a hinged trailing edge surface, such as a wing with an aileron, or a horizontal stabilizer with an elevator. With the trailing surface down (1), a tailwind will cause the airfoil to move downward. With the trailing surface up (2), a tailwind will cause the airfoil to move upward.*

Low visibility cautions

Low visibility while taxiing is not a frequent problem for the VFR pilot, but it is a problem on occasion. For the IFR pilot, it happens much more often. Fog, for example, can obscure one section of an airport and other sections will be in the clear. Airfields near large bodies of water are susceptible to sea or lake fog. It's not unusual for one or more runways to be in the clear, permitting VFR operations, while the parking ramp and taxiways are immersed in a thick fog.

At an uncontrolled airport, you have to be very careful not to run into another airplane or vehicles; at a controlled airport, the folks in the tower won't be able to see ground traffic unless the airport is equipped with ground radar, which is an expensive luxury only the largest and busiest airports have.

Blowing snow is a frequent visibility problem in the winter at some airports. Even on clear days, the wind might whip up fine, dry snow, causing localized areas of poor visibilities. Propwash from other airplanes and rotorwash from helicopters often create near whiteout conditions.

Even the propwash from your airplane can cause snow to swirl up and around the outside of the cockpit, making it seem as though you're taxiing in a personal little snow cloud. If this happens, the trick is to taxi a little faster; this will usually bring you out in front of the propwash cloud. Of course, slick pavement might make a fast taxi an uncomfortable solution to the visibility problem, in which case it's probably better to taxi slower and accept the reduced visibility. With blowing snow, you also have the hazard of snow drifting and obscuring taxiway and runway markings, signs, and lights, so extra care is needed.

Night taxiing requires additional vigilance, as well. It's quite easy to become disoriented, even at well-lit airports. It's helpful to have a copy of the airport diagram handy so you can refer to it easily. Take your time and taxi at a slower speed than you use during the day. If you are unfamiliar with an airport, ask ground control or the tower for assistance. At most airports, there's usually less traffic at night and the controllers are more than willing to give directions.

At nontower airports, you're on your own. Obviously, you should use taxiways if they're available, instead of taxiing on the runway. Even with the landing and taxi lights on, aircraft in the pattern might have difficulty seeing you. If the airport doesn't have taxiways, taxi on the right side of the runway, keep the lights on, and be alert for other traffic visually and by monitoring the appropriate radio frequency.

Some pilots make it a habit to taxi the full length of uncontrolled airports at night to make sure there aren't any unexpected obstructions. In some areas, animals frequently graze on turf runways or find the radiant warmth of paved runways comfortable. During hunting season, deer and other game might have learned that the local airstrip is a safe haven.

More taxi tips—any time

If you're unfamiliar with the airport, study the airport diagram before taxiing or landing and keep it handy in the cockpit to use as a reference.

Read back clearances to cross runways and taxiways and use your full callsign.

Look before crossing or entering any runway to ensure that no traffic is coming your way. If there is any doubt which runway you are paralleling, confirm the proper heading with the heading indicator.

While taxing, check the proper operation of the turn indicator by making shallow S-turns. Of course, pilots of taildraggers should be making continuous series of S-turns to make sure the area ahead is clear.

Use lights whenever possible to be clearly seen. Position lights and anticollision beacons are only required between sunrise and sunset, but it's wise to use them during the day, too; however, strobe lights can be very distracting to other pilots when not far away.

ENGINE RUN-UP CHECKS

The possibility of encountering adverse weather during your flight should make no difference to the care you take when performing the before-takeoff engine checks. These checks should be done just as carefully before every flight no matter what the weather.

Position the airplane near the takeoff end of the runway, out of the way of other aircraft. Point the nose as nearly as possible into the wind to minimize engine overheating, but in a position that you can see other traffic around you. Also consider other aircraft, people, and structures that are behind you. These parameters are often conflicting so you might have to compromise with a slight crosswind and some restricted view.

Set the parking brake.

The normal engine checks include the magnetos, the carburetor heat, and general engine operation. Smoothly move the throttle forward to increase engine RPM to the figure specified in the pilot's operating handbook or checklist. By this time the engine should be warmed up enough so that the engine accelerates smoothly and the oil temperature and pressure indications are in the normal range.

Magneto check

Check the magnetos by switching one magneto OFF (RIGHT or LEFT), note the RPM drop, then switch to BOTH magnetos ON, note the RPM increase, then repeat with the unchecked magneto. Be honest about checking RPM with one magneto off. Was it really within the required range? If you have doubts, do the check again. Sometimes the cause of a low RPM reading is a fouled spark plug; you can often remedy this by leaning the mixture slightly and running the engine for a few minutes. Go back to full rich, and try the magneto check again. If there's no change or you still have doubts after the second check, have it looked at by a mechanic. When finished, double-check both magnetos ON.

Propeller check

If the airplane has a variable-pitch propeller, move the pitch control from HIGH RPM to LOW RPM and back again, according to the pilot operating handbook. If the weather is cold, it's a good idea to do this a few times to make sure the oil is circulating properly.

Carburetor heat check

When checking the carburetor heat, pay attention to the outside air temperature and humidity. It's possible the carburetor iced up while taxiing or even in the short time it takes to do the magneto check. Carburetor ice can occur without visible moisture at temperatures that are higher than 90°F, but it is most likely in humid air at temperatures below 60°F. It's also likely to occur at low power settings; therefore, don't just pull the heat on, note the RPM drop, then push it off. Leave the heat on for a few moments and watch the RPM. If RPM drops initially then increases after 10–20 seconds, you probably picked up some carb ice while taxiing. If a lot of ice has accumulated, the engine might run rough until all the ice is melted. Leave the heat on a little longer to ensure all the ice is gone, then do the check again.

Normally, carb heat should not be used during taxi because the air from the carburetor heat system, in most airplanes, is not filtered. Unfiltered air could cause dirt to get into the engine cylinders and cause excessive wear to the cylinder walls and piston rings.

FINAL PRETAKEOFF CHECKS

Follow the checklist for your airplane to ensure that the following are checked before you taxi to the runway for takeoff.

Flight controls

- Check for freedom of movement and proper operation of ailerons, elevator, and pedals.
- Check operation of flaps and set for takeoff.
- Set elevator trim for takeoff.

Fuel

- Set to the desired fuel tank, usually the one that has the most fuel.
- Switch on the fuel boost pump, if equipped, and check proper operation.
- Mixture control to rich.

Flight instruments

- Check artificial horizon position and adjust if necessary.
- Set in proper barometric pressure in barometric altimeter (altimeter setting). Check that the altitude it gives corresponds to field elevation.
- Check heading indicator and compass.

Radios

- Proper frequencies set in comm radios.
- Proper frequencies set in navigation equipment. Systems indicating proper position.
- Transponder ON and squawking proper code (1200 for VFR or as otherwise instructed by ATC).

Other

- Seat belts fastened and secure.
- Windows and doors secure.
- Parking brake off when ready to taxi to the runway.

TAKEOFF

I bet you thought we'd never get here. Your two biggest concerns during takeoff will be wind and runway surface conditions (Fig. 2-6). Wind, of course, can work for you or against you. Any surface conditions besides dry and hard will make things more difficult.

The hazards from wind come in a number of forms. Most often the hazards are crosswinds. Ever since the pioneering days of aviation, pilots have known that taking off directly into the wind is the best way to do it. In fact, the earliest airfields were simply open fields without marked runways. Pilots could take off or land directly into the wind, no matter what its direction.

Fig. 2-6. *After completing all the preflight steps, it's finally time to take off. Your primary concerns during takeoff are wind and runway surface conditions. Be sure to clear for other traffic, too, particularly at airfields without control towers.*

Unfortunately, there are few airfields today that permit such operations. Most airports, even those with grass strips, have designated runways. Airport designers have taken pains to align runways with the predominant winds in the area, but predominant does not mean always. Crosswind operations are common and sometimes it's necessary to take off and land with a tailwind. Sometimes it happens inadvertently. Try to use the runway that gives the most headwind and the least crosswind component.

Tailwinds

One thing many new pilots don't realize is that a tailwind has a greater detrimental effect on takeoff distance than a headwind has a positive effect. I know it seems logical that the improvement in takeoff performance from a headwind would equal the deterioration in performance from a tailwind, but this is not the case. Look at the takeoff distances in Table 2-1 for a hypothetical light single-engine airplane with a liftoff speed of 60 knots and a no-wind takeoff distance of 1,800 feet.

As the wind increases, you lose more by taking off with a tailwind than you gain by taking off into a headwind. For example, taking off into 25 knots of wind, the airplane would use only one-third of it's no-wind takeoff distance; however, taking off with 25 knots of tailwind, the same airplane would use slightly more than twice it's no-wind takeoff distance. Of course, this isn't like a bank where what you save on one day, you can spend on the next. The runway you don't use taking off into a headwind today, you can't apply to the extra runway you need when taking off with a tailwind to-row. That would be great if it were possible, but unfortunately it doesn't work that way.

Table 2-1. Effects of headwinds and tailwinds on takeoff distance

Wind speed	Takeoff distance Headwind	Change	Tailwind	Change
0 kt	1,800 feet		1,800 feet	
5 kt	1,512 feet	–16%	2,111 feet	+17%
10 kt	1,247 feet	–31%	2,448 feet	+36%
15 kt	1,012 feet	–44%	2,808 feet	+56%
20 kt	801 feet	–56%	3,186 feet	+77%
25 kt	612 feet	–66%	3,618 feet	+101%

For the mathematically gifted, the formula for determining a headwind effect on takeoff roll is $Sw = So(1 - Vw/V)^2$. The formula for determining a tailwind effect on takeoff roll is $Sw = So(1 + Vw/V)^2$. Notice the only difference between the two formulas is the change of the sign preceding "Vw/V." In the formulas, "Sw" is the takeoff distance corrected for wind; "So" is the no-wind takeoff distance; "Vw" is the wind velocity; and "V" is takeoff airspeed.

At larger airports with airline traffic, various considerations sometimes dictate accepting an active runway with a tailwind. (It could be, for example, the only runway with an ILS approach.) Depending on traffic conditions, you might have to accept a tailwind departure or end up waiting a long time. Runway length is the obvious concern, but if you have 10,000 feet of runway in front of you, a takeoff with a 25-knot tailwind in the hypothetical single-engine airplane is certainly hypothetically possible.

Most of the time no more than a 10- or 15-knot tailwind is tolerated by the airport authorities until the active runway is changed to one with a headwind. I wouldn't recommend accepting more than a 15-knot tailwind in a light single-engine airplane, no matter what the runway length. If nothing else, the takeoff run is going to be very different from what you're used to. You'll pass a lot of real estate as the airplane accelerates to liftoff speed and it's just not going to feel right, even if it is hypothetically possible.

Crosswinds

With respect to crosswinds, it must be remembered that each airplane has limits. To be certified under FAR Part 23 (the regulation that most factory-built light general aviation aircraft are certified under), an airplane must demonstrate no tendency to ground loop in 90° crosswinds with velocities up to 0.2 times V_{SO} (the stall speed or minimum steady flight speed in landing configuration). For an airplane with a stall speed of 50 knots, that's 10 knots of crosswind, which is not very much; however, it is possible for the demonstrated crosswind component for a particular airplane to be more than 0.2 times V_{SO}. For example, 0.2 times V_{SO} for the Beechcraft C23 Sundowner is 10.2 knots, but it has a demonstrated crosswind component of 17 knots.

To determine the crosswind component of the wind for a particular runway, you need to know what the wind is. Surface wind information is available from numerous sources, including the tower at controlled airports, flight service, the ATIS or AWOS/ASOS broadcasts, and unicom operators. At noncontrolled airports, the local FBO might relay wind information, although many will not do this any more because of liability concerns. You might have to guesstimate the wind using the wind sock and an ATIS or forecast for a nearby controlled airport. If you operate frequently from airports without surface wind reports, a hand-held wind gauge would be a good investment.

Fortunately, surface wind direction is given in relation to magnetic north so there's no need to apply magnetic variation when comparing wind direction to runway heading. (Remember that winds aloft reports give wind direction oriented to true north.) Runways are oriented toward magnetic north, but rounded up or down to the nearest 10°. In other words, the centerline of Runway 25 could actually be aligned with the magnetic azimuths from 245–254°.

To determine the headwind and crosswind components of a given wind, you can use a wind component chart or table, a copy of which is included in most pilot operating handbooks and on many E6-B flight computers. For quick calculations, use Table 2-2 to figure the approximate crosswind component of the surface wind.

Table 2-2. Determining the crosswind component

Degrees off runway heading	Crosswind component
0–10	None
15	25% of total wind velocity
30	50% of total wind velocity
45	75% of total wind velocity
60	90% of total wind velocity
90	100% of total wind velocity

Although there are two basic methods for landing with a crosswind—using a crab or slip—and a continual debate among pilots about which one is the better, there is only one procedure for taking off in a crosswind and it is generally accepted, with some variations for different airplanes.

In light crosswinds, line up the airplane on the centerline; for stronger winds, you might want to line up on the downwind side of the runway to provide more surface in case the airplane turns into the wind. Use ailerons to keep the wings level or banked into the wind; like taxiing, upwind aileron up—controls turned into the wind. The rudder is used to keep the airplane tracking straight down the runway. Full up-aileron should be used at the start of the takeoff roll and then gradually reduced to keep the wings level as airspeed increases and the wings become more effective.

Left crosswinds are more difficult for single-engine airplanes because engine torque and P factor already cause the airplane to yaw left.

Liftoff in a tricycle-gear airplane should be done at a few knots higher airspeed than normal to prevent the airplane from settling back on the runway with the nose cocked into the wind; however, don't hold the airplane on the ground too long because too much pressure on the nosewheel could induce wheelbarrowing. Just get enough extra airspeed to ensure a positive climb after takeoff.

Tailwheel airplanes should be held on all three wheels until liftoff, although not everyone agrees with this technique. As soon as the airplane lifts off, it should be allowed to turn into the wind to establish a crab angle to keep it on centerline track during the climb out. A crab into the wind will also provide better climb performance than trying to maintain track by slipping, which is by definition uncoordinated flight.

Crosswind landings are covered in chapter 4.

WINDS TO AVOID

Thunderstorms and wake turbulence are definite adverse conditions you don't want to subject your airplane to during takeoff—or any time, for that matter.

Thunderstorms

Chapter 5 covers thunderstorms in more detail, but it's important to mention their effect on takeoff. You obviously should not take off in the middle of a thunderstorm and probably wouldn't want to anyway. In fact, the best place for your airplane when a thunderstorm hits is in a hangar and the best place for the pilot is inside, drinking coffee. The second-best place for the airplane is tied down securely.

Thunderstorms can grow quickly, but usually there's at least some information about them in the weather forecasts. In the daytime, you can see them coming. At night, don't fly VFR when thunderstorms are forecast unless you have an instrument rating, are flying an instrument-equipped airplane, and have weather radar.

When you see a thunderstorm coming toward the airport before you take off, the safest alternative is to postpone the flight and secure your airplane. If you're caught in the open with the engine running, the best bet will be to turn the airplane into the prevailing wind and try to "fly" out the storm on the ground. You don't want to take off unless you can get off well ahead of the storm and fly a course away from it. One danger of any thunderstorm is the gusting wind outside the rainshaft. And the wind associated with a thunderstorm sometimes comes in the form of a microburst.

Microbursts

A microburst forms when cool dense air descends from a mature or dissipating cloud (Fig. 2-7). After the column of air hits the ground, it swirls up, creating a hori-

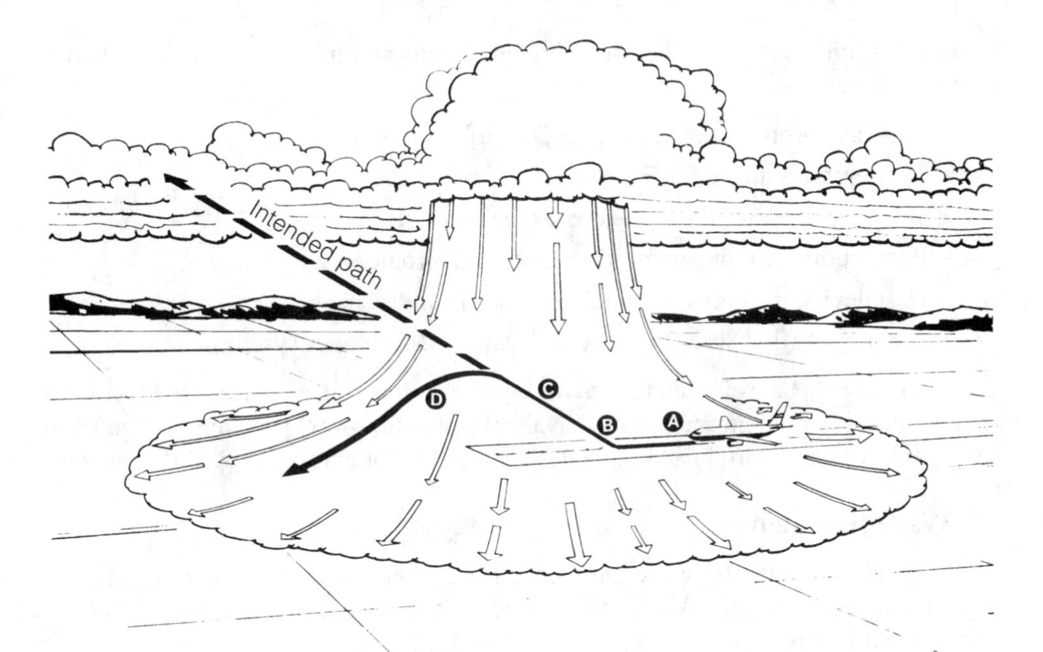

A. Aircraft encounters an increasing headwind.
B. Normal rotation and liftoff.
C. Aircraft starts losing the headwind and begins picking up a downdraft.
D. Aircraft encounters a tailwind, and airspeed decreases dramatically.

Fig. 2-7. *Although microbursts can happen anywhere within the atmosphere, they are most threatening to pilots when flying below 500 feet when there is little time or air space for recovery. Be particularly alert for microbursts when taking off and landing.*

zontal vortex. Think of a stream of water coming out of a hose pointed straight down at the ground. The microburst continues to intensify for about 5 minutes, then begins to dissipate and is gone in about 10–20 minutes.

Microbursts can be either wet, meaning they occur with rain all the way to the surface, or dry, meaning that the rain evaporates before reaching the ground. Possible wet microbursts are easily identified in daylight by a column of rain under a mature cumulonimbus cloud. Usually the base of the cloud is near the surface. It's estimated that only about five percent of all thunderstorms produce microbursts, but it's nearly impossible to tell if a particular column of rain contains one or not. The best action is avoidance.

Dry microbursts are harder to pinpoint, but the presence of virga under a cloud is a good indication. *Virga* is the misty precipitation that extends down from a cloud, but evaporates and does not reach the ground. Thunderstorms and cumulus clouds with bases higher than 10,000 feet AGL are dry microburst-makers. Avoid taking off, flying, and landing near areas of virga.

The FAA has identified the following conditions and indications of possible microburst activity:

- Thunderstorms, building cumulus, or virga activity.
- Temperatures above 80°F.
- Temperature/dew point spread in excess of 30°F.
- Pilot reports of rapid airspeed changes and turbulence.
- Dust devils, blowing dust, and gusty wind conditions.
- A report of wind shear from a low-level windshear alert system.

If you have to take off with the possibility of encountering a microburst, choose the longest runway available and use max available takeoff power. Maintain gear and flap settings until clear of terrain and obstacles and be alert for abrupt airspeed fluctuations.

Wake turbulence

You should be alert for wake turbulence from other aircraft while taxiing, but it is a greater concern prior to takeoff. The *Airman's Information Manual* describes wake turbulence this way:

"When lift is generated by a fixed-wing aircraft, the pressure differential between the air flowing over the top and over the bottom of the wing creates a roll-up wake effect behind the wing tips. This wake consists of two counter-rotating vortices, the strength of which are determined by the weight, speed, and wing shape of the aircraft. The greatest vortex strength is present when the aircraft is heavy, is slow, and has flaps and gear retracted. Helicopters in forward flight also create sizable vortices and should be accorded the same concern as fixed-wing aircraft in flight."

The weight of an aircraft is the biggest factor affecting wake turbulence because the strength of the vortex increases proportionately with the aircraft weight. *Rule of thumb:* If the other aircraft is bigger than yours, avoid its wake turbulence.

The wake forms behind an aircraft and then descends and spreads out laterally, perpendicular to the flight path. The wake does not form until an airplane lifts off from the ground and it stops after the wheels touch the ground. In light winds (less than 5 knots), the wake can persist a long time after an airplane departs. Stronger winds will usually dissipate the wake vortices relatively fast. In a light crosswind condition, one of the wakes could be blown over the runway and stay there for many minutes; therefore, be alert for wake turbulence when taking off from a runway parallel to another runway where large aircraft are operating.

When taking off behind a departing airplane, note where that airplane rotates, then you should lift off before reaching that point and climb in an area that is above the flight path of the other airplane. This will put you above the wake vortices by the time you reach that point (Fig. 2-8).

When taking off behind a landing aircraft, note where the airplane touches down and plan to lift off past that point. This will put you well in front of the wake (Fig. 2-9).

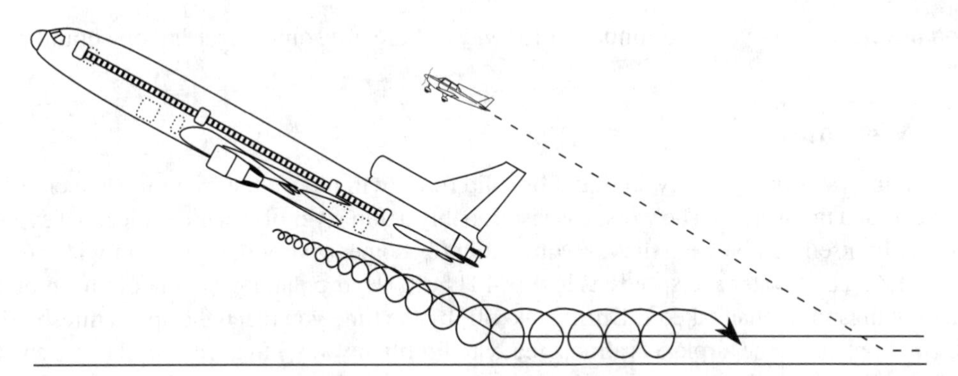

Fig. 2-8. *When taking off behind a departing aircraft, note where that airplane rotates and plan your liftoff point before reaching that point so that you climb above that airplane's flight path.*

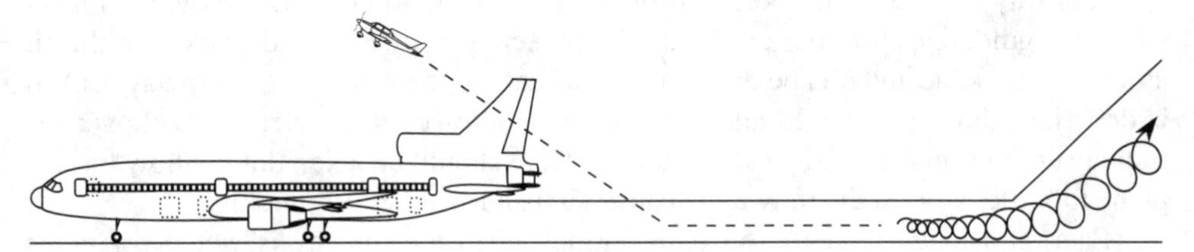

Fig. 2-9. *When taking off behind a landing aircraft, note where the airplane touches down and plan to lift off past that point.*

ATC will provide the following wake turbulence separation to departing aircraft. You may request additional separation; or you may for less separation if you don't consider wake turbulence a hazard:

- Two-minute interval (4- to 5-nautical mile separation) when departing behind a heavy jet from the same threshold, on a crossing runway, or from a parallel runway when staggered ahead of the adjacent runway by less than 500 feet and the runways are separated by less than 2,500 feet.
- Three-minute interval when departing from an intersection of the same runway behind a departing large airplane or when departing in the opposite direction on the same runway behind a large airplane takeoff or missed approach.

Whether or not a warning has been given by a controller, the pilot is expected to adjust the aircraft operations and flight path as necessary to avoid serious wake encounters.

RUNWAY CONCERNS

If you only fly on dry, sunny days and use only paved runways, runway condition is one variable you won't have to worry about much. But if you ever take off or land on a wet, snow-covered, or unpaved runways, there are some hazards you should be aware of.

Wet runways

Any water on a runway, in liquid or solid form, will reduce the friction between the wheels and the surface. The worst case is probably a thin layer of water over ice, but even water by itself can be very slick. Hydroplaning becomes a very real problem with only about ¼ inch of water and speeds as low as 40 knots. (Hydroplaning is more often a problem with landing than takeoff. During takeoff, by the time you reach the speed threshold of hydroplaning, the airplane's almost ready to lift off anyway.) In a way similar to a seaplane lifting onto the *step*—immediately prior to the hull or floats breaking free of the water's surface—hydroplaning might even make the takeoff roll a little shorter, but don't count on it. Standing water encountered by the wheels before hydroplaning would have slowed acceleration and therefore increased the takeoff distance.

One way to reduce the takeoff roll is to do a soft-field takeoff. Follow the manufacturer's guidelines for your airplane. The objective is to get the airplane into the air as soon as possible and escape the retarding effects of the water on the runway. Liftoff is done just above stall speed and the airplane is allowed to accelerate level over the runway until normal climb speed is attained. Care should be taken not to allow the airplane to settle back on the runway because airspeed will be lost quickly.

Plan on a longer takeoff roll than normal when the runway is wet. How much longer is going to require an educated guess. Most general aviation pilot operating manuals don't provide figures for wet runways. As a rule, a wet runway without puddles probably won't require a much longer takeoff roll than on a dry runway. A runway with many pools of water will require significantly more distance.

Another concern is splashed water into a pitot tube or static source. If the temperature is close to freezing or you plan to climb above the freezing level, splashed water could cause erroneous flight instrument indications, frozen wheels and brakes, and frozen control surfaces.

Snow- and ice-covered runways

Snow-packed runways with temperatures well below freezing provide surprisingly good traction. Be alert for icy patches caused by melting and refreezing. Most of the time, operating on snow-packed surfaces is not much different from dry surfaces—if taxi speeds are kept lower than normal and you take your time. Pay attention to snow berms and rutted areas and don't slam on the brakes (Fig. 2-10). Be particularly wary if it is windy. The combination of a stiff crosswind, a sloping taxiway or ramp, and a slick surface could easily result in the airplane sliding one way while you want to go the other (Fig. 2-11).

The most hazardous runway conditions occur when the temperature hovers near the freezing point. Slushy, wet snow is even worse than standing water for slowing down a takeoff roll and increasing the takeoff distance. It also tends to stick to wheels

Fig. 2-10. *Ice- and snow-covered taxiway and runway surfaces require extra vigilance. Take particular care while turning, when the combination of a slick surface, centrifugal force, and a wind gust could send your airplane sliding.*

Fig. 2-11. *Be alert for snow* berms *(piles of plowed or drifted snow), especially along taxiways, when operating a low-wing airplane.*

and other airplane surfaces much better, increasing the probability of problems after takeoff if it freezes.

A small amount of slush on a taxiway is tolerable if you taxi through it slowly, but slush on the runway dictates a no-go decision until it's cleared off. In fact, the FAA recommends that takeoffs in turbojet aircraft should not be attempted when standing water, slush, or wet snow greater than ½ inch in depth covers an appreciable part of the runway. Tests with turbojet-powered airplanes have proven that ½ inch of slush adds 15 percent to the takeoff distance, 1 inch of slush adds 50 percent to the takeoff distance, and 1¼ inch of slush doubles the takeoff distance (Table 2-3). Although these figures apply to turbojet aircraft and therefore cannot be directly applied to piston-powered aircraft, it is undeniable that water, slush, and snow on a runway surface retards acceleration and increases takeoff distance for all aircraft.

Table 2-3. Effect of slush on takeoff distance (turbojet aircraft)

Slush depth	Takeoff distance increased
0.50 in.	15 percent
1.00 in.	50 percent
1.25 in.	100 percent
2.00 in.	Takeoff not possible

If the runway is snow-covered and the temperature rises to above freezing, you can end up with a layer of water on top of ice. Freezing rain can create a similar condition on a formerly dry, cold surface. These are the absolute worst runway conditions. The coefficient of friction is almost zero. The slightest slope becomes impassable and it's almost impossible to simply stand up. Here's a handy rule of thumb: If you can't walk without falling, don't take off—definite no-go conditions.

Unpaved runways

Gravel taxiways and runways require special care (Fig. 2-12). To improve traction in soft areas, special balloon-like tundra tires work best, but you can achieve similar results by reducing the tire pressure in normal tires. Slow engine-idle speeds and minimum-power taxis should be used to prevent throwing gravel into the propeller, fuselage, and wings. If a smooth, gravel-free surface can't be found to do the engine runup checks, it might be preferable to do them during the initial stages of the takeoff roll with plenty of stopping distance if something is amiss. (This is an exception to the general rule about paying full attention to aircraft control.)

Soft-field airstrips are notorious for having nonuniform surfaces because some spots are firmer than others. This can be a problem for pilots used to the paved airports. Generally, the center of runways and taxiways is firmer than the edges; however, be alert for signs of erosion that might indicate softer areas anywhere on the airport surface. Parking ramps might be partially paved, but the paved area is probably for the use of commercial operators. At most airports, you'll be allowed to park there if it's vacant, but will be expected to move if a commuter airline arrives.

To avoid getting bogged down in soft areas, it might be necessary to keep the airplane rolling when you get it started. This way you only need to apply the power

Fig. 2-12. *Gravel strips are notorious for causing damage to props, flaps, landing gear, and tail sections. If a smooth, gravel-free surface can't be found to do the engine run-up checks, it might be preferable to do them during the initial stages of the takeoff roll.*

needed to start moving from a dead stop one time. On a soft surface, you might need nearly full power and this will obviously throw a lot of small rocks into the airplane—the fewer times you have to do this the better. Taxi speed should be enough to keep the airplane moving, usually about a fast walk. Taxiing faster could cause problems if a wheel bogs down in a soft spot.

If you do have to stop at the end of the runway before takeoff, avoid the temptation to apply power against the brakes. Open the throttle slowly so that you achieve full power at about 20 knots. This will decrease the possibility of propeller vortices lifting up stones and throwing them into the tail.

The use of flaps on takeoff is good for the rest of the airplane, but bad for the flaps, particularly with a low-wing airplane. Unless the runway is very short—right on the limit of what the performance charts say you'll need to take off—try to avoid using flaps in a low-wing airplane. The probability of minor damage is just too great, likely near 100 percent. In a high-wing airplane, the probability isn't as great because the flaps are so much higher off the ground. Don't use a flap setting greater than the manufacturer recommends for soft-field takeoffs; if the runway is long enough, it might not be necessary to lower the flaps at all (Fig. 2-13).

Fig. 2-13. *If the runway is long enough, you may get away from using flaps on takeoff and thereby avoid the possible damage caused by the wheels kicking up gravel.*

Maintaining a straight path during takeoff is much more crucial on gravel than on a paved surface because swerving can skid stones onto the airplane. Holding back pressure on the yoke will take weight off the nosewheel and further reduce stone spray. As airspeed picks up, return the yoke to the normal takeoff position.

After takeoff, some pilots recommend holding the airplane level with the runway to increase airspeed and clean small bits of gravel from the wheel wells before retracting the wheels. I don't know how effective this is, but it might be worth a try if you don't have obstacles at the end of the runway to contend with.

AFTER TAKEOFF

Shortly after takeoff, activate your VFR flight plan by giving your departure time to the nearest flight service station (Fig. 2-14). You can request that the tower do this for you if you're operating from a controlled airport.

Do the after-takeoff checklist, set the climb power, check the time and note it on the flight planning log, and turn to the first heading or turn as prescribed by ATC.

You're on your way.

Fig. 2-14. *After takeoff from an uncontrolled airport, you can activate your VFR flight plan by calling the nearest flight service station.*

3
En route hazard strategies

"These veterans were to be seen in the field restaurant—gruff, not particularly approachable, and inclined somewhat to condescension when giving us the benefit of their experience. When one of them landed, rain-soaked and behind schedule, from Alicante or Casablanca, and one of us asked humble questions about his flight, the very curtness of his replies on these tempestuous days was matter enough out of which to build a fabulous world filled with snares and pitfalls, with cliffs suddenly looming out of fog and whirling air-currents of a strength to uproot cedars. Black dragons guarded the mouths of the valleys and clusters of lightning crowned the crests—for our elders were always at some pains to feed our reverence. But from time to time one or another of them, eternally to be revered, would fail to come back."

Antoine de Saint-Exupery, *Wind, Sand, & Stars*

THE WHOLE POINT OF PLANNING AND PREFLIGHTING IS TO REDUCE RISK SO that you do come back, unlike some of those veterans that Saint-Exupery revered so much. In flying, you can never reduce risk to zero. Most pilots probably wouldn't want to do that anyway. When there's no risk, there's no challenge and flying without the challenge is as exciting as sitting in your living room watching a video about flying—or sitting as a passenger on an airliner. So you have to accept some risk or you'll never get off the ground.

But there's no sense taking unnecessary risks. When you take off, you will have evaluated numerous risks and considered them low enough to give you a high probability of completing the flight as planned. If, on the other hand, you feel the risks are such that the probability of completing the flight is not high, then you shouldn't take off.

Icing, thunderstorms, low visibilities, mountainous terrain, power lines, human errors, emergencies: Anything and everything can happen en route. All of part 2 is about en route hazards; here are six strategies to help you cope with them.

THE FIRST STRATEGY

Disregarding environmental effects for the moment, the chances of an engine failure or other mechanical problem necessitating an emergency landing are the same regardless of the terrain you fly over. Why, then, do we say flying over some terrain, such as mountains or large bodies of water, entails more risk?

It's riskier because the chances of surviving an emergency landing are less; therefore, to reduce the risk, we need to prepare for an off-airport landing. Carrying survival equipment appropriate for the terrain we fly over will not make the engine run any smoother, but it will give us the advantage if we find ourselves spending a few hours or even days far from civilization.

The first strategy for reducing the risks of en route hazards is to be prepared for an off-airport landing. This harps back to chapter 1, namely planning before the flight. Planning for the "unplanned-for-landing" is so basic that it's often overlooked. If pilots overlook it, then how much more likely is it that our passengers won't even think of it.

For most passengers, the difference between a trip in an airplane and a trip in an automobile is that the airplane is faster. Their primary concern, with respect to what they should wear and take along, is with their destination; little or no thought is given to terrain they'll be flying over. As a pilot, it's your job to make sure you and your passengers are prepared to spend a cold night in the mountains or a wet night in a life raft or simply an extra night in a hotel room. Being properly clothed and having survival gear available after an emergency landing can make the difference between life and death (Fig. 3-1).

Being prepared for a precautionary off-airport landing changes the pilot's mental attitude. A precautionary landing is made because you decide it's more hazardous to continue flying than it is to land during the next few minutes. An emergency landing is made immediately, such as after an engine failure in a single-engine airplane; you must land because you don't have any other choice.

With a precautionary landing you have the choice of flying or not flying. The choice might be between flying through turbulent air in a mountain pass or landing at

Fig. 3-1. *Being properly clothed for the terrain you plan to overfly could make the difference between surviving a few uncomfortable hours in the wilderness after an emergency landing and not surviving.*

a small strip at the mouth of the pass to wait for better conditions. The choice might be continuing the flight with an odd engine noise or landing in an open field. When it's necessary to make decisions like these, you don't need to complicate the decision-making process with concerns about how warmly you and your passengers are dressed or if you're all going to be uncomfortable if you can't get back into the air before nightfall or if you will be plagued by mosquitoes or if you'll all go hungry. In other words, your decision regarding flight safety should not be unduly tainted by your concern about what will happen after you're on the ground.

I say "unduly tainted" because the consequences of any off-airport landing should always be considered when you're confronted with such a situation. Landing in a field or on a road *is* hazardous and the possibility of damage or injury must be factored into the equation when contemplating a precautionary landing; however, by carrying survival equipment and being properly dressed, you give yourself more options.

For example, remote, unattended landing strips become acceptable precautionary landing sites when you know you'll be able to survive a few hours or days until help arrives. If you fear for your survival after making the landing at a remote strip, you will be more likely to decide to continue the flight and could end up having to make an emergency landing in an even less favorable area.

THE SECOND STRATEGY

"Until help arrives" brings up the second strategy in reducing the risk of en route hazards: the VFR flight plan. Filing a flight plan is step 6 in chapter 1. According to the *Airman's Information Manual*, "A filed flight plan is the most timely and effective indicator that an aircraft is overdue." The manual also quotes from the National Search and Rescue Plan: "The life expectancy of an injured survivor decreases as much as 80 percent during the first 24 hours, while the chances of survival of uninjured survivors rapidly diminishes after the first three days." An Air Force review of 325 search and rescue missions revealed that an average of "36 hours passes before family concern initiates" an alert.

When filing a VFR flight plan, it's best to file only one leg at a time. If you file a lengthy flight plan with many stops, a search will not be started until 30 minutes after the ETA for the final stop, unless other information is received. The initial search is not by aircraft, but by radio and telephone. From experience, rescue coordination centers and flight service stations know that many more pilots are delayed or forget to close their flight plans soon after landing than are actually overdue and missing. It might take another hour or so before a rescue aircraft actually starts looking for you.

If you don't file a flight plan, at least tell someone where you're going and when you plan to arrive. Call them when you get there. Don't count on being able to get out a Mayday call if you run into trouble. You might not have time to do it. You should try to do it, of course, but don't count on being able to and don't count on somebody hearing it, particularly at low altitudes and over remote areas.

Remember, filing a flight plan is the most inexpensive insurance you can buy—it's free.

THE THIRD STRATEGY

Although it dovetails well with a VFR flight plan, VFR flight following is an entirely separate strategy. The term flight following is a carry-over from the time when flight service stations across the country would monitor an aircraft's progress via radio contacts if the pilot wrote "flight following requested" in the remarks section of the VFR flight plan form. Today, because of extensive radar coverage, the closing of many flight service stations, and the proliferation of onboard transponders, the service has been taken over by ARTCC controllers. The proper term is *radar advisories* or *traffic advisories*, but controllers will still know what you mean if you mention flight following with the formal terminology.

The new way of flight following has advantages and disadvantages when compared to the old way. It's better because the exact position of your aircraft is known to someone on the ground as long as you stay within radar coverage. If you have an emergency, ATC will know where to send search and rescue (Fig. 3-2). With the old flight following, your position was less precise and a search was not initiated until after you failed to report at your next checkpoint.

Fig. 3-2. *By filing a flight plan and requesting traffic advisories or flight following from ATC, you increase the probability that you will be found quickly by search and rescue. (U.S. Air Force Sikorsky HH-3E rescue helicopter.)*

The primary disadvantage of the new system is that traffic advisories are not always available to the VFR pilot. The main purpose of ATC is to separate IFR traffic; therefore, if the controller for a particular area is too busy, she might refuse your requests for traffic advisories. Or, if she does accept the request, she might not pay much attention to you, and might even forget you if she becomes busy with other aircraft; however, if you did have an emergency, you could easily gain her attention.

Despite the fact that you might be ignored during busy times, being under radar coverage while VFR is better than being completely on your own. In reality, you still are very much on your own. It's still your responsibility to see and avoid other traffic, just like it's the responsibility of pilots on IFR flight plans to do the same when in visual conditions. It's still your responsibility to maintain VFR visibility and distance-from-clouds minimums. And it's still your responsibility to navigate, maintain obstacle and terrain clearance, and avoid special-use airspace. You will get traffic advisories when the controller is able, but be aware she might miss some conflicts and it is still your responsibility for the safe conduct of the flight.

Your responsibility to the controller, after you have requested traffic advisories, is to pay attention to the radio chatter and listen up for your call sign. If you want to switch frequencies to talk to flight watch (*see* the fourth strategy) or unicom or whomever, advise the controller before leaving her frequency. If you want to change your cruising altitude or route of flight, let her know. Technically, you don't need her approval, but there just might be a good reason, such as conflicting traffic, which would make it good for her to know what you're doing. The controller might even be able to provide you with a suggested heading to your destination or limited advisories about inclement weather.

The main reason for traffic advisories is to have someone to call if you have a problem. Sure, you can always try the emergency frequency (121.5) and set the transponder to the emergency setting (7700), but if you can simply key your microphone and talk to a controller who already knows your position, you'll be one step ahead of the game. She might even be able to recommend a nearby airport for an emergency landing. That could make the difference between destroying your airplane in a bumpy pasture and making a dead-stick landing to a runway.

An added advantage of using flight following is that it fulfills the requirement to establish radio communication with ATC prior to entering Class C airspace. It will also help when approaching Class B airspace because the controllers will know you're coming and what you plan to do.

THE FOURTH STRATEGY

Make productive use of your time while en route. Few pilots talk about it, but most will admit that the time spent en route often becomes monotonous and boring (Fig. 3-3). With modern navigation equipment and radar control, there is really very little for the pilot to do besides monitor her instruments, check her fuel burn, and get updates on the weather. These things hardly take any time at all. The old description of flying being hours and hours of boredom punctuated by moments of stark terror is just as true today as it was decades ago—except that the moments of stark terror are becoming fewer and fewer.

But just because you don't *have* to do as much, doesn't mean there's nothing to do. In fact, there are a number of things you can do that will not only help pass the time en route, but will also make the flight more enjoyable and safer, probably the most important issue.

First of all, you can continually improve your position awareness by following along on an aeronautical chart and cross-checking your position with several navaids. Don't rely only on your VOR, GPS, or loran receiver. Look out the windows and try to

Fig. 3-3. *The en route phase of flight often becomes monotonous when everything is going well, even over the most beautiful terrain. Use the time productively by following your track on a chart, staying current with the weather, reviewing aircraft systems and emergency procedures, and monitoring the instruments.*

identify prominent and not-so-prominent features by using your map. When you positively identify your position on the map, find a nearby VOR compass rose, estimate the radial, and then check it with your aircraft's equipment. You can also look for airfields along your route, in case you need to make a precautionary or emergency landing. Try to know where the closest airport is at all times, without using the nearest-airport feature on your loran or GPS.

Doing this every flight has the obvious side benefit of improving your map-reading ability. Maybe you'll never need to use it as your sole source of navigation, but you never know. Passengers also usually appreciate it when you can name a town or river when they ask because the quick response gives them much more confidence in your ability as a pilot.

When airports are few and far between, look for suitable emergency landing sites. Ask yourself, "What if the engine stops right now? Where should I land?" Make choices while everything is working properly so that the decision will be quicker if something does happen. Refer to chapter 9 for the relative merits of different landing surfaces.

On the subject of emergencies, pull out the pilot's operating manual and go through the checklist for each malfunction. Take your time and picture in your mind what indications you'll see or hear. Then, as you read each item in the checklist, put your hand on the appropriate switch and simulate the required action. If you review each emergency this way at least once a month, you'll be well prepared when something does happen.

Quiz yourself about aircraft systems. A handy way to do this is to look at each switch, control, and instrument in the cockpit and explain the function of the item and the system that it controls or monitors as if you were having to do this for an FAA examiner for a type checkride. If you stumble during the explanation, pull out the pilot's operating manual and read about the system. Don't forget to cover limitations associated with the system.

Are you on an IFR flight plan or do you anticipate having to file IFR and make an approach at your destination? Use the time en route to review the approach charts for the airport. Look at all the approaches, not just the one you expect or hope to get. Become familiar with all the approaches the airport has to offer; you never know when a change in wind, an equipment outage, or a traffic problem might call for a different approach on very short notice, perhaps in rapidly deteriorating weather conditions.

Finally, every 10 or 15 minutes, take a good look at the instruments and ask yourself, "What's wrong with this picture?" Look hard to see that each switch is in the proper position. Check the fuel selector for the proper tank. Check fuel burn. Is fuel consumption what you expected when you did your flight planning? What about the communication and navigation radios? Do you have the proper frequencies selected? If the navcom unit has active and standby frequency positions, do you have the next frequencies that you'll need on standby?

Far from being boring, you can use your time en route for productive activity. You can keep ahead of the aircraft, stay aware of your position, and brush up on your aircraft knowledge.

THE FIFTH STRATEGY

Become friends with EFAS, the en route flight advisory service that is better known as flight watch. Referencing to the *Airman's Information Manual* again: "EFAS is a service specifically designed to provide en route aircraft with timely and meaningful weather advisories pertinent to the type of flight intended, route of flight, and altitude. In conjunction with this service, EFAS is also a central collection and distribution point for pilot reported weather information."

In other words, a call to flight watch will get you the latest up-to-the-minute weather information about your route and the latest pilot weather reports, which often provide the best information you can get. Your only cost is a pilot report to flight watch as a courtesy to the weather reporting system; this is not an absolute requirement. "Pilots are encouraged to report good weather as well as bad, and to confirm expected conditions as well as unexpected to EFAS facilities," according to the AIM.

AIM also states: "EFAS is not intended to be used for filing or closing of flight plans, position reporting, getting complete preflight briefings, or obtaining random weather reports and forecasts." In other words, don't ask New York Flight Watch for the terminal forecast for Miami because you plan to hop on a Miami-bound airliner at Newark after landing your small plane at Linden Airport. Flight watch will provide you with destination weather and terminal forecast on request, but en route flight advisories are meant only for the phase of flight that begins after climb-out and ends with descent to land.

Making periodic calls to flight watch is also a good way to keep track of severe weather forecast alerts, sigmets, convective sigmets, and airmets. All EFAS facilities have equipment to directly access radar displays from National Weather Service radar sites.

To contact EFAS, the radio call should prefix flight watch with the name of the air route traffic control center identification serving the area of your location and follow it with your aircraft identification and the name of the nearest VOR to your position. This approximate location is needed so the specialist can select the best transmitter/receiver for communication with you.

Charts depicting the location of the flight watch facilities are in the *Airport/Facility Directory*, which you should have onboard as a result of proper preflight planning. Prior to the flight, review the EFAS chart and make a note or two on your flight log regarding the appropriate ARTCC. If you don't know which flight watch area you're in, simply say, "Flight watch, (your aircraft ID), (nearest VOR)," and the specialist will respond with the name of the facility.

The common flight watch frequency is 122.0 MHz and is available from 6 a.m. to 10 p.m. local time. Coverage is assured for aircraft flying from 5,000–17,500 feet AGL, but you can usually obtain contact with a facility at a lower altitude.

Recall that pilot weather reports are often the most beneficial weather information you can receive. It's unfortunate that more pilots don't provide more pireps more often. In some weather conditions, such as icing, pireps provide the most timely and accurate information. The more rapidly conditions change, the greater the need for pireps.

Air traffic facilities are required to solicit pireps when ceilings are at or below 5,000 feet, visibility is at or below 5 miles, thunderstorms are in the vicinity, light or heavier icing is observed, moderate or heavier turbulence is observed, and wind shear is encountered. Pilots are urged to volunteer reports of these conditions promptly whenever observed. If there isn't time to send a report while in flight, a pirep can be phoned into the nearest FSS or weather service office after landing.

The standard pirep format is listed in Table 3-1. It would be handy to have a copy of this with you in the cockpit at all times. Every time you make a call to flight watch, make it a habit to provide a pirep. This is not a requirement, but if every pilot did it one time each flight, we would improve weather information for all of us.

When making a pirep, remember that a ceiling is, by definition, the height *above ground level* of the lowest layer of clouds or obscuring phenomenon which is reported as *broken*, *overcast*, or *obscuration*, and is not classified as thin or partial. Sometimes it's very obvious what the ceiling is, but at other times it can be very difficult to judge.

Table 3-1. Standard pilot weather report (pirep) format

1. Aircraft identification.
2. Position.
3. Time—Coordinated Universal Time (UTC).
4. Altitude—in feet MSL.
5. Type aircraft.
6. Sky cover—ceiling height and cloud tops in feet MSL and coverage (scattered, broken, or overcast).
7. Weather—flight visibility, precipitation, and restrictions to visibility (haze, smoke, dust, etc.).
8. Temperature—in degrees Celsius.
9. Wind direction—in degrees magnetic and speed in knots.
10. Turbulence—see turbulence reporting criteria table.
11. Icing—see icing definitions.
12. Remarks—to report additional elements or clarify previously reported items.

Pilots are not expected to be qualified weather observers; they should just report what they see. It's also anticipated that pilots will use their barometric altimeters to determine cloud heights; therefore, pilot-reported cloud *heights* will actually be cloud *altitudes*, because they are expressed in feet above mean sea level. It's not necessary to try to determine feet above the ground when making pireps.

Also, with respect to reporting winds at altitude, it's likewise anticipated that pilots will report wind direction in degrees magnetic, as opposed to degrees true as is done in winds aloft reports; therefore, it's not necessary to convert the wind direction indicated by a loran or GPS receiver to true north before making a pirep.

THE SIXTH STRATEGY

You call EFAS to get updates on the weather. Knowledge about hazardous weather is useless if you don't act on it. So the final strategy is to divert toward the good weather (Fig. 3-4).

This might sound like very simplistic advice, but it's much harder to follow in practice than it seems. The purpose of most flights is to get to a specific destination, not just anywhere. The myriad of forces pulling you to that destination will be awesome. The desire to arrive is sometimes called "get-there-itis" or "get-home-itis," but the *need* to get to that specific place is often rooted deeper than the mere desire to simply be there.

A destination is a goal. The achievement of goals is a prime motivator in our society. Learning to fly and obtaining your pilot's license were probably two of your big goals since childhood. The can-do attitude is highly respected in life. In the military if you're not gung ho or if you don't press on at 110 percent, you're considered a slacker, which is the worst possible thing you can be. To achieve one's goals at all costs is considered admirable by many, if not most, people.

Cessna Aircraft Company

Fig. 3-4. *Make it your personal policy to head for the good weather whenever you find out that the weather at your planned destination has turned bad.*

In aviation, however, achieving goals at all costs takes on a more serious meaning than it does in the corporate boardroom. Risking your life's savings on a new business venture is entirely different from risking your life trying to make it to your destination. This might be surprising to some, but the primary goal of any flight should not be to arrive at the destination, but rather to complete the flight safely.

Getting to a destination often is the primary motivation for making the flight in the first place, but it should not become the primary goal of the flight. When it becomes the primary goal, it often does so at the expense of safety. Accident statistics are filled of examples of what happens when getting there becomes more important than flying safely.

If this still seems contrary to conventional wisdom, think of it this way: If you ask any passenger on board any aircraft—private, corporate, commuter, airline, medevac, or military—if they would rather land safely 100 miles from the planned destination or crash 1 mile from that destination, what do you think the answer will be? I'd be willing to bet my life's savings that the unanimous choice will be the former.

Making the goal flying safely instead of reaching the destination requires a different mental attitude on the part of the pilot and this attitude must also be conveyed to your passengers. When we travel by any form of transportation, we do so because we want to get somewhere. Because we travel so much by ground transportation, we are accustomed to arriving, more or less on time, at the destination. Only under very rare circumstances is it necessary for us to alter our travel plans because of weather or other conditions. This is so unusual that when it does happen, we usually talk about it for years afterwards.

But flying is different, particularly general aviation flying. Our priorities must be rearranged. Getting there is not the most important thing. In decreasing order of importance, Table 3-2 reflects what your priorities should be.

Table 3-2. The right priorities

1. Safety of passengers and crew.
2. Safety of the aircraft.
3. Completion of the mission.

Actually, you probably already apply these same priorities when traveling by car, although you might never have thought about it this way. When you drive, you're undoubtedly concerned about the safety of your passengers, your car, and yourself, but the actions required to minimize driving hazards are habitual and automatic; therefore, you focus on getting to the destination. Rarely is it necessary to divert from your planned route and it's even less rare that you have to divert to an alternate destination. (This is not to imply that our desire to reach a destination when driving never gets the better of us. If you've ever nodded off at the wheel, you know what I mean.)

With respect to flying, putting the safety of crew and passengers first means there will be times you will not make it to your planned destination because the weather you encounter en route is simply not good enough. When this happens, your strategy must be to divert to the good weather.

What is good weather? When flying VFR, look for no clouds below 1,000 feet above the highest obstacles along your route and visibilities no less than 3 miles. Practically, this means that if you can't see the ground clearly over the nose while staying 1,000 feet above obstacles, then you should head for better weather or find the closest airport and land.

Remember, too, that any precipitation will lower visibility. So if you're flying right at minimums and see precip ahead, either go around it or turn around before you get there. The loss of daylight will also reduce effective visibility, particularly if there is any obscuration in the air. You'll also lose the ability to see areas of lower visibility (rain) or clouds creeping down.

For an IFR flight, good weather is more relative. Ceiling and visibility should obviously be above minimums for a particular approach. If you're flying to an unfamiliar airport, using higher values as your personal minimums is wise.

A FINAL THOUGHT

Somewhere during my twenty-something years in aviation, I heard or read the following: "Bad weather isn't really dangerous—it's the marginal weather that will kill you." This makes a great deal of sense because when the weather is really bad, the decision to not fly into it is an easy one. When the weather is only marginally bad, the decision becomes a judgment call, and judgment calls are notorious for being swayed by outside factors, like the desire to get to a meeting or home to loved ones.

So think about the above advice when you find yourself unsure of the advisability of continuing toward not-so-great conditions. "The weather's not that bad . . . I think I can make it" is not a positive thought.

Remember, it's the marginal weather that will kill you.

4
Approach and landing

"Takeoffs are optional. Landings are mandatory."

Unknown flight instructor

YOUR FLIGHT IS ALMOST COMPLETE. ALL THAT REMAINS IS THE DESCENT from cruise altitude and the landing. Hopefully the worst of the adverse conditions is behind you, and this last part of the flight will be a piece of cake.

Unfortunately, this is not always the case. After hours en route, just when you feel the need for a break, you must be at your most alert state. Obstacles are a big hazard, particularly at night and when flying IFR. Although instrument approaches are carefully designed to provide minimum obstacle clearance during all phases of the procedure, numerous accidents have occurred when pilots were flying an authorized approach. Obstacles are also a problem for the VFR pilot who finds scud running necessary for the final few miles to the destination airport.

Weather is often a big factor, too. Although you can often avoid bad weather on the way toward your destination by changing your route, you can't do anything about the weather at your destination, except go to an alternate. Because your destination is the place you want to be, you'll be inclined to accept worse weather there than you normally would. After all, what's the point in flying 400–500 miles just to end up somewhere you don't want to be?

Finally, the physical act of landing is one of the most difficult maneuvers that most pilots have to accomplish, particularly crosswind landings. Add to this difficulty numerous visual illusions and it's easy to understand why this final phase of flight is not always the piece of cake we'd like it to be (Fig. 4-1).

Fig. 4-1. *Landing is perhaps the most difficult phase of flight for most pilots. Don't wait until the downwind leg to prepare yourself and your airplane.*

APPROACHING TO LAND

This phase of a flight begins when you start the descent that will eventually lead to the wheels touching the ground. Like the other phases of flight, it requires preparation. The pilot flying IFR will usually have more to do than the pilot flying VFR, but both need to get ready for the upcoming approach and landing.

The flight conditions and weather at the destination airport will obviously dictate if an instrument approach is required or if one can proceed visually; however, if you are flying an IFR flight plan and you're not in a particular hurry to land, fly a full instrument approach, even if the weather is clear and a million. Every pilot can benefit from a practice instrument approach, even professional pilots who do them often. Use a hood or other vision-restricting device if you want—only if you have a safety pilot on board who meets FAR requirements—but even without such a device, you gain valuable experience flying the approach.

Although this book is not primarily about instrument flying, there are a few things that bear mentioning because they apply to IFR and VFR operations.

First, when you are still a good distance from the airport, about 30–40 miles, but perhaps farther depending on local air traffic procedures, find out the current weather at the field. If the airport has an ATIS, AWOS, or ASOS, then listen and write down the current weather. If you have to listen three or four times to copy the whole thing, do it. It's very easy to transpose numbers or forget them completely, and I have always found it very reassuring to look down on the flight log and read the current weather conditions before starting the approach.

If the airport doesn't have a recorded or automated report, then you'll have to call the nearest flight service station or even ask ATC. If you are flying IFR or are on VFR flight following, then be sure to advise the controller you'll be leaving the frequency for a minute. Obtaining the latest weather takes a few minutes, which is why you should do it before you get too close to the airport when you'll probably be much busier with other radio calls.

Most of the time, with the ATIS or other weather information in hand, you'll be able to decide or anticipate your next move. If the weather is good VFR all the way, you could cancel IFR and proceed visually, unless you take my advice and fly a practice approach. On the other hand, if the weather is such that an instrument approach is required, the ATIS broadcast usually will tell you which one is being used. If it doesn't, the wind direction will clue you in to which approach to expect; therefore, you have time to pull out the appropriate approach charts and begin preparing for the approach.

If the weather report tells you the field is close to or below minimums, you will have an early warning that a missed approach is a strong possibility and you can prepare for alternatives. Of course, if the weather is below minimums, it's usually wise to proceed to your alternate from your present position. Only under very rare circumstances is it worth proceeding to the destination with the weather reported below minimums.

For example, if you know for certain that the below-minimums condition is being caused by a small, but intense thunderstorm that is moving quickly over the airport and will clear the field before you arrive, then you could safely proceed. Similarly, if you know that the morning fog that's obscuring the field is likely to burn off before you get there or shortly thereafter and you have enough fuel to hold for an hour or two, it's probably safe to continue toward the destination.

CHECKLISTS

Following a checklist is always important, perhaps most important prior to landing. The following generic checklists apply to most airplanes, but you should, of course, refer to the pilot's operating handbook for the particular airplane you're flying.

Predescent checklist

- Destination weather written down.
- Altimeter set to destination airport barometric reading.
- Seat belts and shoulder harnesses fastened.
- Approach materials ready (when IFR, the proper approach charts; when VFR, the airport diagram).
- Engine, mixture, carburetor heat, and propeller properly adjusted and set as required.
- Flaps as required.
- Fuel tank selection as required.
- Landing gear down. (Be sure to check for proper illumination of the GEAR DOWN indicators; don't just move the handle and assume the landing gear extends.)

Final approach quick check

For most light airplanes the GUMP check is usually sufficient because you are rechecking tasks that presumably have already been accomplished. GUMP stands for:

- Gas (check fuel selection handle)
- Undercarriage (a nice British term for landing gear checked down and locked)
- Mixture (full rich)
- Propeller (variable-pitch propeller at full RPM)

With more complex aircraft, I like to check:

- Warning lights (no lights showing or no new ones that I don't already know about)
- Instruments in the green (all instruments showing normal indications)

Short final check

On short final, I like to check the landing gear down one more time. It's a good habit to get into and has saved my bacon more times than I'd like to admit. Although checking that the landing gear is down is not a concern for pilots who only fly aircraft with fixed gear, it's especially important for pilots who fly both fixed- and retractable-gear airplanes. It's better to check that the wheels are down in a fixed-gear airplane than to forget to put them down in a retractable-gear airplane (Fig. 4-2).

A memory jogger some pilots use is to check that the gear is down when they are told "Cleared to land" by the tower. This works much of the time, especially at busy airports where you often don't receive landing clearance until on short final; however, when you get clearance to land 10 miles out before you have even done the predescent checklist, it's not as effective as a last-minute reminder.

Fig. 4-2. *A quick check on short final that the landing gear is down is a good habit to get into, even if you only fly fixed-gear airplanes. Someday this habit might save your skin if you "move up" to a retractable.*

NONSTANDARD IFR-IN-VMC APPROACHES

Although visual and contact approaches are available only to pilots flying on IFR flight plans, VFR pilots should be familiar with them to know what's going on when they hear another aircraft cleared for a visual or contact approach. Both have the potential of increasing the probability of traffic conflicts, and the contact approach is a particularly good scud-running inducer.

Visual approach

One of ATC's time-savers is the visual approach, as defined in the *Pilot/Controller Glossary*: "Visual approach (is) an approach wherein an aircraft on an IFR flight plan, operating in VFR conditions under the control of an air traffic control facility and having an air traffic control authorization, may proceed to the airport of destination in VFR conditions."

Now this is fine and dandy and is usually a win-win situation for both the pilot and ATC. Commercial pilots particularly like to get visual approaches because it usually cuts a few minutes off the flight time, thereby saving the carrier some money and making the passengers happy. ATC likes it because they can sequence in more aircraft in less time than if each one had to fly an instrument approach.

The tricky part is that the pilot must have the airport in sight or have a preceding aircraft in sight before ATC will issue a clearance. There are three pitfalls here.

First, you have to maintain VFR conditions all the way to the airport, so in addition to simply identifying the airport from 10 miles out, or however far you are, you must also look for clouds or precipitation between yourself and the airfield. This can be particularly hazardous at night, when you might not be able to see clouds below you, but can see the airport.

Second, sometimes positively identifying the preceding aircraft is difficult, if not impossible. There might be an aircraft ahead of you and it might appear to have the same shape and number of engines as what the controller told you it would be, but can you really tell a Boeing 767 from an Airbus 310—both are twin-engine airliners—when you are 5 miles away? I can't.

Third, there is no firm definition of "airport in sight" and consequently pilots and controllers interpret the phrase different ways. To some pilots, it means seeing the entire airport: runways, tower, terminal, hangars, and the like. To others, it means having the approach lights or the rotating beacon in sight. To others, just seeing the general area where they know the airport to be is enough. Also be aware that altitude makes a big difference when trying to locate the airport. A pilot in an aircraft that is 1,000–2,000 feet above your own will invariably report "field in sight" before you do, even if you are a few miles closer. The point is, don't be intimidated to call "airport in sight" and accept a visual approach before you really do have a good ID on the field.

Whenever you're in doubt about the airfield being in sight or in doubt about the identification of a preceding aircraft, it's always safer to continue the approach IFR and not accept a visual approach.

Contact approach

A contact approach is not used often because it is beneficial only under rare circumstances. Before seeing what those circumstances are, let's define a contact approach. According to the *Pilot/Controller Glossary*, "Contact approach (is) an approach wherein an aircraft on an IFR flight plan, having an air traffic control authorization, operating clear of clouds with at least 1 mile flight visibility and a reasonable expectation of continuing to the destination airport in those conditions, may deviate from the instrument approach procedure and proceed to the destination airport by visual reference to the surface. This approach will only be authorized when requested by the pilot and the reported ground visibility at the destination airport is at least 1 statute mile."

Now how might this come in handy? Let's say, for example, that you are approaching an airport with a visibility minimum of 1½ miles, but the reported visibility is only 1 mile. Legally, you may begin the approach and continue to the missed approach point, at which time you must have the runway environment in sight to legally continue past the missed approach point. But with 1-mile visibility, you won't see the runway environment until you are ½ mile past the missed approach point.

But let's say that when you reach the minimum descent altitude you are below the clouds and as you look into the murk you identify a road or other distinguishing landmark that you know will lead you to the runway. So you request a contact approach from ATC. You are still flying on an IFR flight plan, but now you are guiding yourself to the field visually. Of course, if you lose contact with the ground after requesting a contact approach, you will have to execute a missed approach procedure.

If this sounds like scud running to you, I'm not surprised. It sounds like scud running to me, too; however, it would fit the criteria for safe scud running if the following rules are observed (*see* chapter 5 for more on scud running):

- The pilot has at least 200 hours instrument time. (Because you're flying IFR, I assume you have an instrument rating and the aircraft is equipped for IFR.)
- The airplane is in Aircraft Approach Category A, meaning the airplane has an approach speed slower than 90 knots.
- It is daytime.
- No other approach will get you to the airport IFR.
- You know the airport environment and the approach.
- The approach doesn't take you up a blind canyon.

Remember when requesting a contact approach, that you must be on an IFR flight plan, you must have ATC authorization, and you are responsible for your own obstacle clearance.

LANDING ILLUSIONS

The visual cues that pilots use while flying are accepted as being extremely important, but not completely understood. Some experts believe peripheral vision is of primary importance during landing while others say it is only of secondary importance; however, we do know that about 80 percent of the information we obtain is visual. And no one would argue that any restrictions to normal visibility make the pilot's job harder.

Landing is the time we need visual information the most. En route flying straight and level on instruments or even VFR with limited instrumentation in the cockpit is not difficult for most pilots after only a few hours of training. Takeoffs with limited visibility require more skill, but even zero-zero visibility takeoffs are not that difficult if you can manage to keep the wheels tracking along the centerline until you reach flying speed. But landings are something else entirely.

Most commercial, corporate, and military pilots have the luxury of expensive vision-enhancing or automatic-approach equipment to help them get down in bad weather. A Category III ILS takes the largest of airliners down to the runway, with the pilots only watching the gauges and monitoring the function of the automatic system. Head-up displays, for many years the province of only military fighter aircraft, have become standard equipment on airliners and are now showing up in turboprops and corporate aircraft. Forward-looking infrared and night-vision goggles, again developed mainly for the military, will soon find their way into civilian use as a visual approach aid. (They're already being used in police work and search and rescue.)

Eventually, these things, or something even better, will become available at reasonable prices to general aviation, greatly enhancing the capabilities of both VFR and IFR pilots; however, for the time being, most of us will be left with our own vision. What we see is what we get. Unfortunately, what we see is sometimes not exactly what we think we're seeing.

Rain and snow

Rain has a tendency of making things, especially lights, seem farther away. It can also fool depth perception by diffusing the positional relationship of objects, making it difficult to perceive relative distances. Rain on the windshield creates the illusion of greater height.

Falling snow can cause the same effect, although usually not to the same extent as rain. On the other hand, heavily falling snow can restrict visibility almost as effectively as fog, as can blowing snow.

Snow on runways and surrounding areas reduces contrast and can make depth perception worthless. Like other featureless terrain, the absence of ground features can create the illusion that the aircraft is higher than it actually is.

Fog

Besides restricting forward visibility, fog diffuses lights and objects, making it difficult to judge their precise position and distance. Like rain, fog can make objects appear farther away than they actually are. Penetration of fog during descent might create the illusion of pitching up, causing you to steepen the approach abruptly if you don't recognize the illusion.

Runway slope

Runways are rarely level from one end to another, but the slope of most runways is hardly noticeable. There are some, however, with pronounced slopes—either upslope or downslope, depending on the landing direction (Fig. 4-3).

When you approach an upslope runway at the normal glide angle, your view of the runway will cause you to feel that you are coming in too high and your reaction will be to flatten the approach angle.

When you approach a downslope runway at the normal glide angle, your view of the runway will cause you to feel that you are coming in too shallow and your reaction will be to steepen the approach angle.

The solution is to use the VASI or PLASI lights, if installed at the field.

Terrain slope

Similarly, if the terrain under the approach path is not level, you might misjudge the proper approach path.

If the ground rises up to the threshold of the runway, you will feel that you are too high on the approach and you might decrease your height too soon.

If the ground descends toward the threshold, you will feel that you are too low on the approach and you might come in too high.

Runway width

A narrower-than-normal runway can create the illusion that the aircraft is higher than it actually is (Fig. 4-4). The pilot who does not recognize this illusion will fly a

Fig. 4-3. *The downhill slope of this "one-way" airport in Alaska is quite noticeable from ground level. Approaching from the air, however, it might not be as evident. Downhill-sloping runways can give the illusion that you are coming in too shallow, causing you to steepen the approach angle unnecessarily.*

Fig. 4-4. *An unusually long runway will appear narrower and therefore create the illusion that the aircraft is higher than it actually is. This might cause you to fly a lower approach, with the risk of striking objects or landing short.*

lower approach, with the risk of striking objects or landing short. A wider-than-usual runway will have the opposite effect, causing you to feel that you are lower than you actually are.

Flicker vertigo

There is some debate among medical professionals about flicker vertigo, but I know from experience that the rhythmic flashing of the sun through an airplane's propeller or a helicopter's rotor blades can make me feel nauseous.

In an airplane, this can be particularly acute when landing or taking off into the sun when it is low on the horizon. Although this has not happened often during takeoff, I have noticed it a number of times when landing at twilight because the prevailing wind at my home airport is from the west and—obviously—the sun sets in the west. The flickering seems to hit on short final and during the flare, just at the time that I need a good view of the ground the most.

The only way to stop flicker vertigo is to not look at the light source through the rotating element. Turn your head away as soon as you notice the flickering—the sooner the better. Don't try to gut it out and think that you can force yourself not to be affected. Apparently some people can become involuntarily mesmerized by the flickering light and actually go into a sort of coma. This is not the sort of thing you want to happen to you in flight.

WIND

Although it's surprising at first, it makes sense after you think about it: Of all adverse weather conditions, wind is cited most often as a cause in weather-related accidents of general aviation aircraft, and most of these wind-caused accidents occur during landing (appendix B).

Wind, of course, can be caused by any number of factors (covered in more detail in chapter 5). Often, when we're in the landing phase, there isn't much we can do about the adverse winds other than be aware of them and fly our aircraft accordingly: to the best of our ability. At other times, we can use our judgment and good sense to help avoid or mitigate the ill-effects of ill winds.

For example, at a tower-controlled airport with a selection of runways, you don't have to accept landing with a strong crosswind on the active runway just because it's the active runway. If you feel the crosswind is going to exceed or even come close to your ability to handle it, request a landing clearance to another runway that gives you less of a crosswind. You might have to fly an extended pattern or make a few 360° turns because of other traffic, but the extra time is well worth it if you avoid bending metal on landing. The airport is there for you to use safely; do not feel forced to use a certain runway when you know it would be safer to use another one.

In a similar vein, at nontower, uncontrolled airports, you have the right to use the runway that gives you the best wind conditions. Just be sure you make your intentions known and that you pay attention to other traffic (Fig. 4-5). As often happens at un-

Fig. 4-5. *At uncontrolled airports, you may choose the runway direction that you feel is safest for you to use. Just be sure to inform other aircraft at the field what you are doing and then await your turn.*

controlled airports, people get into the habit of using the same runway, even when the wind would make the use of another runway better. All it takes is one person to say, "Hey, why don't we switch to Runway 27 today?" and everyone else gladly follows along, thankful that one soul spoke up.

Other wind conditions that are often avoidable are thunderstorms, microbursts, and wake turbulence. (Thunderstorms are also briefly discussed in chapter 2 and fully discussed in chapter 5; microbursts and wake turbulence are also discussed in chapter 2.) Although you can't always avoid these hazardous conditions, you usually will have some warning of their presence. The safe thing to do is to give all three a wide berth. The problem is you often can't tell how bad a thunderstorm is until you get into it, and then, if it is really bad, it's too late; accept the fact that any thunderstorm should be avoided if possible.

The following hangar story, told to me by my father, Ralph C. Padfield, well illustrates the perils of underestimating the power of an approaching thunderstorm prior to takeoff and subsequently having to land as it hits the airfield.

Hangar story

It was Easter Sunday 1945. I was a U.S. Naval Aviation cadet in primary flight training at Memphis Naval Air Station, Tennessee. As I marched back from breakfast with the other cadets in my squadron, I noticed the sky was dark with low overcast. I felt sure flying would be canceled for the day. I don't know why we were scheduled to train on Easter Sunday in the first place, although it might have been because the Navy was short of aviators, the battles of Iwo Jima and Okinawa having taken their toll earlier that year. As it turned out, all the squadrons but mine were dismissed from training. We marched off toward the flight line.

We were training in bright yellow Stearman biplanes designated N2S by the Navy and affectionately called the "Yellow Peril" by the pilots (Fig. 4-6). I was scheduled for a solo practice flight, and as I did my walkaround inspection I kept glancing at the recall light on the tower because the sky was getting darker and the clouds lower. We relied on visual signals from the tower because there were no radios in the airplanes. The light remained green and there was no red "Baker" on the recall mast.

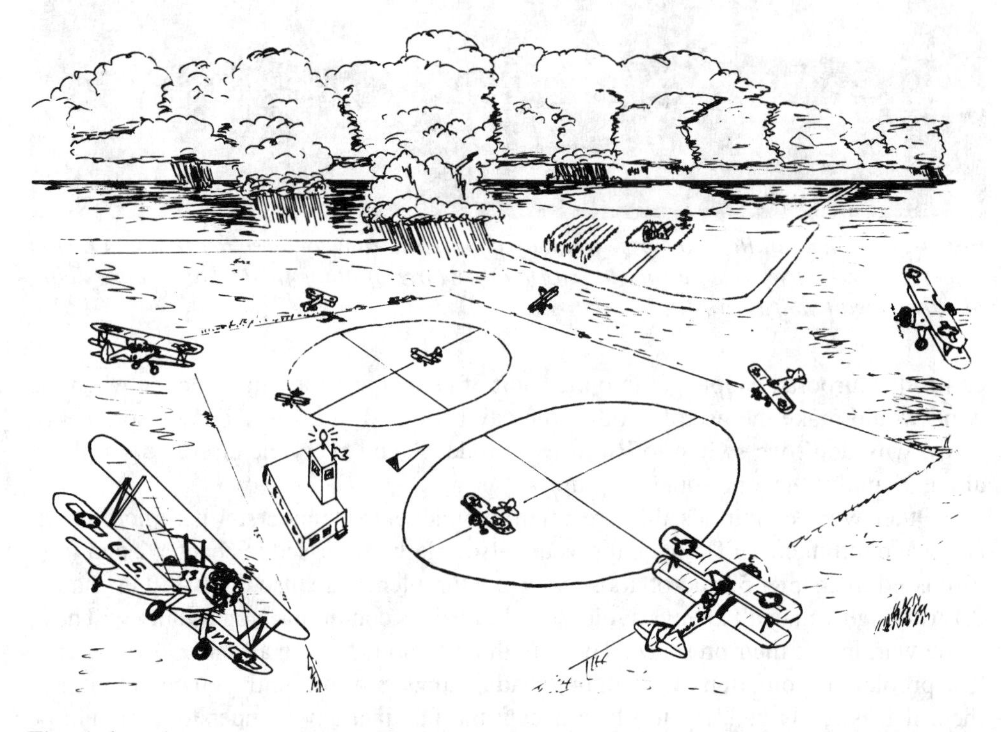

Fig. 4-6. *"The storm was now almost over the field, and as I watched, the recall light changed to red and the Baker flag was raised quickly to the top of the mast.... Suddenly, both mats looked like targets for a multitude of yellow dragon flies all trying to land at once."*

Meanwhile, a tar-mac [attendant] arrived and inserted the hand crank in the inertia starter. After I climbed into the cockpit, he called, "Switch off," and began turning the prop to get gas to the cylinders. By the time I had settled myself and my parachute into the bucket seat and secured the shoulder straps and seat belt, the tar-mac had cranked the inertia starter to a typical high-pitched whine.

When he yelled, "Switch on!" I flipped the switch and pulled the starter handle. With a cough and sputter, the engine caught and the prop wound up to a steady idle.

After the tar-mac pulled the chocks, I eased the throttle forward and fell into the fish-tailing line of Yellow Perils heading to the takeoff area.

Instead of runways, NAS Memphis had two circular blacktop mats painted with spoke-like wind lines. The two mats lay about a quarter mile apart on either side of the control tower. A wind tetrahedron, also located between the mats, pointed to the takeoff and landing direction. As my line of planes approached the mat, each one picked an open wind line and proceeded to take off into the wind. I noticed the overcast was still becoming darker and more threatening, but the ceiling was still above minimum.

I took off and leveled at 500 feet AGL, staying in the traffic pattern until the tower was in view. One glance showed the recall light still green, so I continued in the pattern toward the heading for my practice area. To the west, I could see a thunderstorm just entering the area. Lightning flashes were already visible.

Just before reaching my planned exit point, I looked back at the tower. The storm was now almost on the field, and as I watched, the recall light changed to red and the Baker flag was raised quickly to the top of the mast. I was probably the first to see the recall signals, but being upwind, I had to stay in the traffic circle and wait until I was farther downwind before I prepared for landing. By this time, rain was drenching the wings, but thanks to the relative wind, no rain came into the open cockpit.

When my left wingtip was abeam the end of the mat, I cut power and turned for the landing approach. Suddenly, both mats looked like targets for a multitude of yellow dragonflies—all trying to land at once. Luckily, there were only two planes ahead of me and I quickly lined up on one of the wind lines.

As my airplane slowed and approached the ground, the wind became gusty and the rain increased. I had trouble keeping the upwind wings down and also stay lined up for landing. I decided a wheel landing would be safer than the standard three-point carrier-type landing, so I pushed the nose down and increased the throttle. My speed increased, but the gusts seemed more manageable.

Within seconds, I placed the wheels on the mat and gradually reduced speed. As the airplane slowed, rain water poured into the cockpit. Lightning was now flashing on all sides. With no delay, I aimed toward a waving tar-mac who guided me to a pair of chocks. Throttle off, switch off, and a dash across the taxiway brought me to the ready room—and warmth.

Those of us safely on the ground watched in fascination as yellow biplanes landed on the two mats. I saw one touch down a bit too tentatively and begin to ground loop, but the pilot was able to recover and finish landing safely. We found out later that others weren't so lucky.

Several airplanes were caught at the practice fields and were forced to land. Some were damaged, but most remained flyable and were able to return to the station later that day after the storm had passed. Luckily, no one was seriously hurt. It was obvious to everyone that the powers that be in the tower had clearly underestimated the intensity of the approaching weather system until it was much too late.

Wake turbulence

The same rule of thumb concerning wake turbulence during takeoff applies for landings: If the other aircraft is bigger than yours, avoid its wake turbulence.

Remember that the wake forms behind an aircraft and then descends and spreads out laterally perpendicular to the flight path. The wake does not form until an airplane lifts off from the ground and it stops after the wheels touch the ground. In light winds (less than 5 knots), the wake can persist a long time after an airplane departs. Stronger winds will usually dissipate the wake vortices relatively quickly. In a light crosswind condition, one of the wakes could be blown over the runway and stay there for many minutes; therefore, be alert for wake turbulence when you are landing on a runway that is parallel to another runway where large aircraft are operating.

When landing behind a heavy aircraft, stay at or above the aircraft's final approach path (Fig. 4-7). Watch where its wheels touch down and land beyond that point. If parallel runways are in use, request the upward runway if practical. If not, stay above the other aircraft's final approach path, note its touchdown point, and land beyond a point abeam its touchdown spot.

When landing behind a departing heavy aircraft, watch where it rotates and touch down well prior to that point (Fig. 4-8). If the aircraft departs from a crossing runway

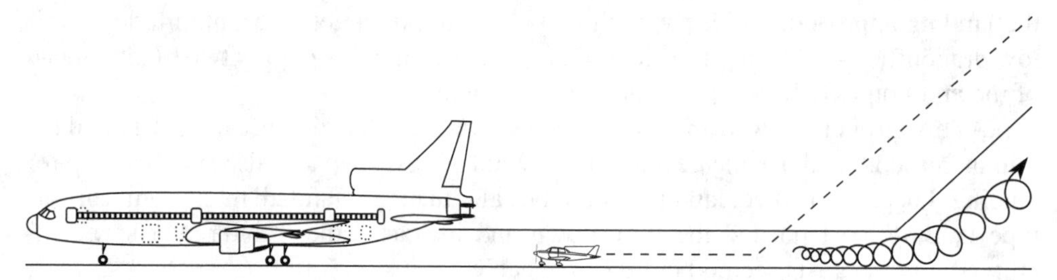

Fig. 4-7. *When landing behind a heavy aircraft, stay at or above the aircraft's final approach path. Watch where its wheels touch down and land beyond that point.*

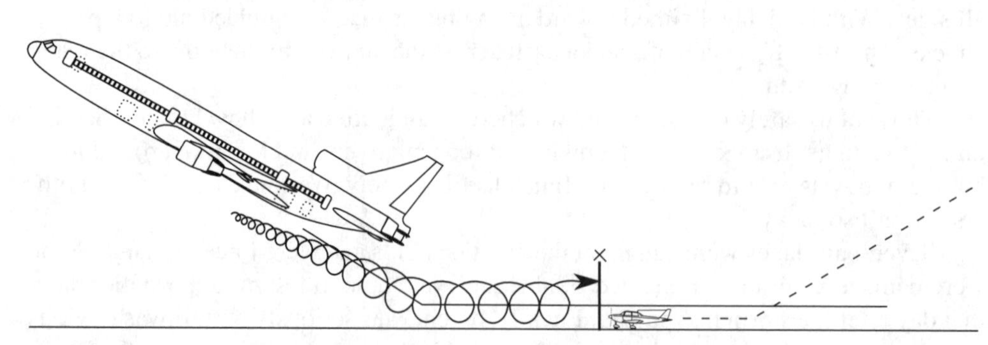

Fig. 4-8. *When landing behind a departing heavy aircraft, watch where it rotates and touch down well prior to that point. Never approach a runway or airport underneath a departing heavy aircraft's flight path.*

and its rotation point is past the intersection, land prior to the intersection. In no instance should you approach underneath the flight path of a departing heavy aircraft.

ATC will provide the following wake turbulence separation to landing aircraft:

- Small aircraft landing behind a heavy jet: 6 nautical miles.
- Small aircraft landing behind a large aircraft: 4 nautical miles.

You may request additional separation if desired; however, whether or not a warning has been given by a controller, the pilot is expected to adjust aircraft operations and the flight path as necessary to avoid serious wake encounters.

Crosswind landings

Two crosswind landing techniques are *crab* and *slip*. From talking to pilots and reading numerous magazine articles (and subsequent letters to the editor), I have concluded that whichever method I favor, there will be experienced, 30-year instructor pilots who will vehemently disagree with me.

First, one thing that all pilots agree on (at least I haven't heard any dissenters) is that the airplane should be landed with the airplane's longitudinal axis aligned with the airplane's direction of movement—in other words, with the wheels tracking directly down the runway. The point is to avoid excessive side loads on the landing gear.

The tricky part is to get the airplane into this position when you have a crosswind.

Two points. First, the limiting crosswind for any airplane is 30 percent of stall speed. This means that a Cessna 172, with a stall speed of 55 knots, should not be landed with a crosswind component of more than 16.5 knots (55 knots × 0.30 = 16.5 knots).

Second, most pilots agree that you should use a slightly higher approach speed when landing with a crosswind. An old rule of thumb says use your normal approach speed plus one-half the maximum wind gust.

Okay. We know we should track a straight line down final approach and there are two ways to do this.

Sideslip technique. With the sideslip crosswind landing technique, which is used by many modern jet transports when making automatic landings, the upwind wing is held down and the nose is aligned with the runway using the rudder. If, for example, the wind is from the right, you hold the right wing down by turning the yoke to the right and hold the nose straight by using pressure on the left rudder pedal. The opposite controls are used with a wind from the left. Because you are using the opposite rudder than you would when making a coordinated turn, the control position is called *cross-controlled*.

Cross-controlled sounds ominously dangerous, and it is if you are trying to enter a spin, but at a normal approach speed it is perfectly fine. The only aerodynamic concern is that you will be descending at a faster rate than with the controls in trim. You can counteract this with a little extra power.

The advantage of the sideslip technique is that you can align the airplane on final and hold the controls in more or less the same position all the way through the flare (you still have to flare by pulling back on the yoke) and touchdown. When done properly, the upwind main wheel touches first, then the downwind wheel, and finally the nose or tailwheel.

Crab technique. The other technique is to crab during the final approach. Crabbing keeps all the controls coordinated and gives a normal descent speed. It looks funny to be approaching the runway with the nose cocked one way or the other, but your track down final will be straight.

The fun starts just before the flare because you want to get the airplane aligned with the runway to avoid the side loads on the landing gear. The only way to do this is to use the sideslip technique. You can do it just before you start the flare, while you are doing the flare, or just after the flare. The crab technique is therefore a combination of crab and sideslip.

Personally, I've found that putting in a sideslip after the flare is too late for me. There's so much else going on and, with the nose rising up and blocking out some of the view, I just don't feel comfortable. I have the same problem if I put the sideslip in while doing the flare. Maybe other pilots can cope with dropping a wing, adding some rudder to get the nose straight, and pulling up to lose airspeed all at the same time, but I find I usually can't do it very well.

I opt for sticking in the sideslip just before flaring. This might be too late for some pilots, but it seems to work well enough for me.

You might have already figured out that my preferred crosswind landing method is to crab down final and then put in the sideslip just before flaring, and you would be right. Perhaps you prefer to sideslip all the way down. That's fine if it works for you.

I do feel, however, that my method—which is not really mine but was taught to me—has one advantage over the sideslip technique. In my experience, the wind is rarely constant all the way down final approach. The wind correction that you put in at 700 feet is not going to be the same correction you need at 300 feet or 50 feet. Consequently, you need to make slight, and sometimes not so slight, adjustments at least two or three times on the approach.

I have found it easier to adjust position with a crab than with a sideslip, so I prefer to crab for as long as I can. Because I'm not going to meet the last wind change until I'm about 50 feet above the ground, I figure I might as well wait to put in the sideslip until at that point. It works for me.

RUNWAY CONCERNS

Runway length and the condition of the surface are two major concerns on landing. The pilot's operating handbook gives advice and procedures you should use for your particular aircraft. Often it's wise to check the performance charts to ensure that the runway is long enough. In the same way that it's always a good idea to add 25 or 50 percent extra to the required takeoff distance shown on the chart, it's also smart to add a similar margin for landing.

Short-field landings

The goal is to touch down as close to the beginning of the runway as possible and with a minimum airspeed in order to reduce floating. This requires careful planning and precise control of the airspeed and flight path.

Plan to approach using full flaps (or as recommended in the operating handbook), a somewhat shallower glide angle than normal, and an airspeed about 1.3 times the power-off stalling speed in landing configuration. This will require some power to achieve a stabilized, uniform glide toward a spot just short of the threshold. The slight floating that occurs during the flare will extend the glide past the threshold. Because a go-around with full flaps is difficult in most airplanes, it's usually best to wait until committed to land before lowering the last increment of flaps, especially in high density altitude conditions.

By using power on the approach, landing accuracy is improved. The glide to the spot can be stretched by adding more power and shortened by reducing power. Maintain airspeed with elevator. In gusty or crosswind conditions, use less flap.

As you approach the threshold and are sure the approach can be safely completed, slowly close the throttle and reduce airspeed to just above stall speed as you flare. Touchdown should be at a minimum forward speed, in a nose-high attitude; however, be careful not to pull the nose up too soon or too high. In both cases, you risk entering a stall too high above the ground, striking the tail, or both.

In a taildragger, the aim is to make a three-point landing with the main wheels and the tailwheel meeting the ground at the same time. In a nosewheel airplane, you want to land on the main wheels with the nosewheel off the ground. After the nosewheel lowers to the ground (or all three wheels are on the ground in a taildragger), the rollout can be shortened by holding back on the yoke, applying the brakes firmly, and retracting the flaps.

Soft, rough, and snow-covered runways

Soft, rough, muddy, and snow-covered runways have the effect of retarding the rollout. The airplane can become mired or stop suddenly enough to nose over or damage the landing gear (Fig. 4-9).

Make the approach like any other full-flap approach until the flare. At this point, the nose should be gradually raised to slow the airplane almost to a stall just above the runway, while adding power to stop the descent. Touchdown should occur at minimum forward and downward speed. A certain amount of power should be left on after touchdown. If the airplane has tricycle gear, the nose should be held up to protect the nose strut from hard impacts against rough or soft spots. Flaps should be kept down to keep as much weight as possible on the wheels. Because they end up in this attitude naturally, taildraggers are inherently better on soft and rough surfaces than nosewheel airplanes, which is why taildraggers are so popular among bush pilots in Canada and Alaska.

Fig. 4-9. *Be alert for unusual conditions on the runway at all times. Drifted snow partially covers one end of the runway at this uncontrolled airport.*

Ice-covered runways

The approach should be made as if to a short field, with the realization that braking action will be limited and it might be difficult to slow down after touchdown. At controlled airports, a report of braking action is usually available; if the runway has enough braking action to safely slow down and stop large aircraft, the runway should have enough braking action for light singles and twins. At uncontrolled airfields, you're on your own, so give yourself a wide margin of error and use the longest runway available.

Crosswind landings to a slick surface are particularly hazardous and should be avoided whenever possible. After touchdown, you might find it's impossible to keep the airplane from being blown sideways, sliding across the runway as airspeed slows and the rudder and ailerons lose effectiveness.

TAXIING IN

The airplane has decelerated and you've rolled to the cross taxiway to turn off toward the ramp. The thunderstorms you passed en route are long gone. Rain is pelting the airplane, but you knew the raindrops on the windshield made it appear you were higher than you really were on the approach and you compensated nicely.

The half-inch of slush on the runway makes going slow but you carefully follow the taxiway lights using the airport map on your lap. The low-lying clouds on the western horizon have reduced the twilight and you know it will soon be dark. You're thankful the flight is over.

But don't relax your guard yet. As Yogi Berra said about baseball, "It ain't over 'til it's over." Pay attention to what's happening outside the cockpit. Take your time and don't rush yourself. Your flight isn't over until your aircraft is securely tied down or hangared and you've closed the flight plan (Fig. 4-10).

Only then you can relax.

Fig. 4-10. *The flight ain't over 'til it's over. Don't let down your guard until your airplane is safely hangared or tied down. (And don't be as blasé about the propeller as this pilot is!)*

5
Weather hazards

"Everybody talks about the weather, but nobody does anything about it."

Charles Dudley Warner, editorial in the *Hartford Courant*, 1897

WARNER'S STATEMENT IS NOT QUITE AS TRUE AS IT WAS 100 YEARS AGO. Cloud seeding, for example, is one area in which people are doing something about the weather. But for most of us, there is little that we can do to purposely change the weather; however, for those of us who are pilots, there is certainly much that we must do *about* the weather.

Unfortunately, some pilots apparently are not giving as much attention to weather as they perhaps should. The latest statistics from the National Transportation Safety Board show that about 25 percent of all aviation accidents are weather-related.

The heating and cooling of air—and its resultant movement over the earth as wind and turbulence—form the basis of all weather. In addition to wind and turbulence, icing, thunderstorms, and lightning are weather phenomena that are all pertinent to flight operations. And finally, continued visual flight into instrument weather conditions is perhaps the most insidious weather hazard of all.

WINDS AND TURBULENCE

> "If fluids were such that turbulence could not occur . . . climate would be very different from what it is now. The atmosphere gets most of its heat from below, and in such an imaginary world the air would be intensely hot in daytime near the ground and very cold at the breathing level of a human being, for there would be no means by which heat could spread rapidly upwards. Evaporation would virtually cease because the film of saturated vapor over a water surface would not be rapidly removed, and human beings and animals could not congregate without fear of being poisoned by their own products."
>
> **Sir Graham Sutton, *Understanding Weather***

The next time you are bumping along in flight, trying to keep the wings level and the nose on the horizon, and watching the vertical speed indicator as it jumps back and forth from plus to minus like a pendulum on a clock, think about the above quote from Sir Graham Sutton, a noted British meteorologist. Maybe it will help you feel a little better knowing that if it weren't for that turbulence "human beings and animals could not congregate without fear of being poisoned by their own products." (What a lovely way to put it!)

Wind is the pilot's friend and foe. With a headwind down the runway, takeoff performance improves and landing rolls are shorter. With a tailwind en route, groundspeed increases and we arrive at our destination earlier than expected. On the other hand, crosswind takeoffs and landings tax our abilities to the very limits and a strong headwind en route causes fuel reserves to dwindle. Light turbulence makes a flight uncomfortable; extreme turbulence can destroy an aircraft. Yes, wind is friend and foe.

Where wind comes from

On the global scale, wind is caused by the uneven heating of the earth's surface. Different surfaces, such as land and water, absorb heat from the sun at different rates and therefore radiate heat to the overlying air at different rates. The warmer air expands, rises, and becomes less dense than the cooler air, as hot-air balloonists well know. The cooler, denser air is drawn to the ground by gravity, forcing and lifting the warm air upward. The warmer air spreads, cools, and eventually descends. This process is called *convection*. What we call wind is the horizontal flow as air masses of different temperatures push each other around in a convective current.

Because the earth's axis is more or less perpendicular to its orbital plane, the equatorial regions of the earth receive more direct light and heating than the polar regions. Intense heating near the equator raises the air temperature and lowers the density. The dense cold air from the poles flows toward the equator and forces the less dense air aloft. This air then flows to the poles to complete the convective circulation.

If the earth didn't rotate on its axis, the wind would always flow in two layers: a lower layer from the poles to the equator and an upper layer from the equator to the poles. The cold air from the poles would heat up as it moved toward the equator, rise at the equator, and then flow back to the poles at high levels. The circulation would be two gigantic hemispherical convective currents (Fig. 5-1); however, because the earth does rotate, this simple circulation model is greatly distorted.

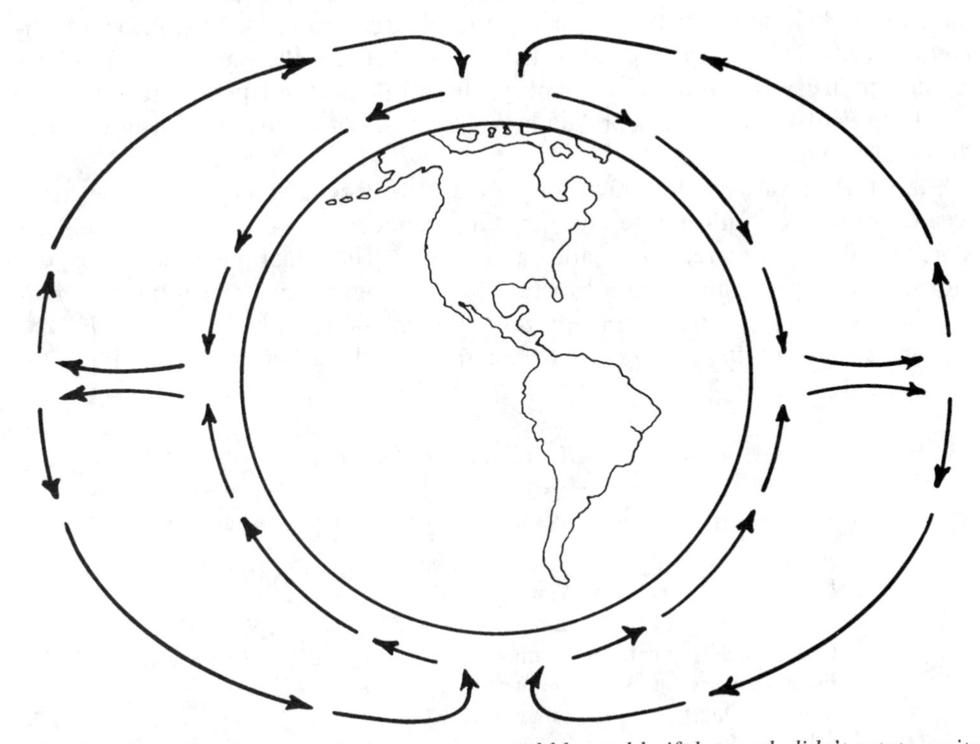

Fig. 5-1. *The global air circulation pattern would be stable if the earth didn't rotate on its axis. A lower layer wind would flow from the poles to the equator and an upper layer would flow from the equator to the poles. As the cold air from the poles flowed toward the equator, it would heat up, eventually rising at the equator to flow back to the poles in the upper layer.*

Rotation in the Northern Hemisphere causes air to flow to the right of its original path; in the Southern Hemisphere, air flows to the left of its normal path. The physical principle that describes this force was first explained in the mid-nineteenth century by Gaspard G. Coriolis, a French civil engineer, and the force now carries his name.

Coriolis force varies with latitude: zero at the equator, maximized at the poles. Wind speed has a bearing: the stronger the wind, the stronger the Coriolis force. The general effect of this phenomenon is to cause the warm air forced aloft at the equator to turn to the right, toward the east, at about 30° latitude in the Northern Hemisphere. The colder air moving southward from the north pole at low level is deflected to the right as well, which means it is moving toward the west. Because winds are named from the direction that they come from (not the direction that they are moving toward), the winds from the equator are called the *high-level subtropical westerlies* and the winds from the north pole are the *low-level polar easterlies*. (The wind directions are opposite in the Southern Hemisphere.)

These deflections of the hemispherical convective currents cause the air to build up at about 30° and 60° latitude in both hemispheres, creating semipermanent high-pressure belts. This disrupts the convective transfer between the equator and the poles, a situation that would prevent the atmosphere from reaching equilibrium if nothing else happened. The impasse is broken by huge masses of air that periodically mix in the middle latitudes.

Air that stays over a surface long enough will eventually acquire the basic weather features—temperature and moisture characteristics—of that surface. The stockpile of air over a region is called an *air mass*. The larger the surface area, the more uniform the air mass characteristics. Meteorologists have categorized six major air mass types: *arctic*, *continental polar*, *continental tropical*, *equatorial*, *maritime polar*, and *maritime tropical*. Characteristics of the different air mass types are in Table 5-1.

Table 5-1. Typical features of air masses

Air mass type	Season	Temperature	Humidity
Arctic	Winter	Very cold	Very dry
Continental Polar	Winter	Cold	Dry
	Summer	Cool	Variable
Continental Tropical	Summer	Hot	Dry
Equatorial	Summer	Very warm	Moist
Maritime Polar	Winter	Mild	Moist
	Summer	Cool	Moist
Maritime Tropical	Winter	Warm	Moist
	Summer	Very warm	Moist

From the north, cold air masses break through and plunge southward toward the tropics. Large storms develop in the middle latitudes and carry warm air northward. From the southern polar regions, cold air masses likewise move northward to the southern tropics. The result is a band of migratory storms with ever changing weather between 30° and 60° in both hemispheres.

The sharp, well-defined boundary between air masses is called a *front*. When a cold air mass follows a warm air mass, a *cold front* occurs. When a warm air mass follows a cold air mass, a warm front occurs. When two air masses stall with no change in position, it is called a *stationary front*. When one front—usually a cold front—moves so rapidly that it overtakes the air mass in front of it and pushes that air mass upward, it is called an *occluded front*. Characteristics of weather changes associated with warm and cold fronts are given in Table 5-2.

Table 5-2. Characteristic weather changes associated with the passage of cold and warm fronts

Weather Element	Cold front Before	Cold front After	Warm front Before	Warm front After
Wind direction	Southwest	Northwest	South	Southwest
Wind speed	Moderate	High	Low-moderate	Moderate
Temperature	Warm	Cold	Cool	Warm
Clouds	Cumulus	Clear	Stratus	Cirrocumulus
Precipitation	Heavy	No	Moderate	Showery
Humidity	High	Low	Moderate	High
Pressure	Low	High	High	Low

This is a general overview of the cause of wind on the large planetary scale. The hows and whys are much more intricate—way beyond the scope of this book. Such things are nice to know, but not really of great relevance to the pilot. The pilot wants to know what the wind will be on takeoff, en route, and on landing. An in-depth knowledge of pressure gradients, Coriolis effect, and other meteorological elements concerning wind is certainly helpful to the person who flies, but not necessary. As important as these effects are to global wind, weather, and climatic patterns, it is often the smaller-scale, more localized causes of wind that are of much greater concern to the pilot.

Some salient points

Before discussing these smaller-scale causes of wind, here are a few things about wind that are good to know.

First, in the Northern Hemisphere the general flow of wind well above the surface of the earth is clockwise around a high pressure area and counter-clockwise around a low pressure area. In the Southern Hemisphere, it is just the opposite. On a weather map showing pressure gradients as contour lines (*isobars*), the wind will be parallel to the contours (Fig. 5-2).

Second, near the surface of the earth, wind is subject to friction. The rougher the terrain, the greater is the frictional effect; therefore, near the surface, the wind crosses the isobars at an angle instead of flowing parallel to them. On a surface pressure chart, wind will be seen spiraling outwards from high-pressure systems and inward toward low-pressure systems.

Third, wind speeds generally increase with height up to the tropopause, about 35,000 feet in North America, where the jet streams are found.

Jet streams

Jet streams are high-speed winds found near the boundary of the two levels of the atmosphere known as the troposphere and the stratosphere (Fig. 5-3).

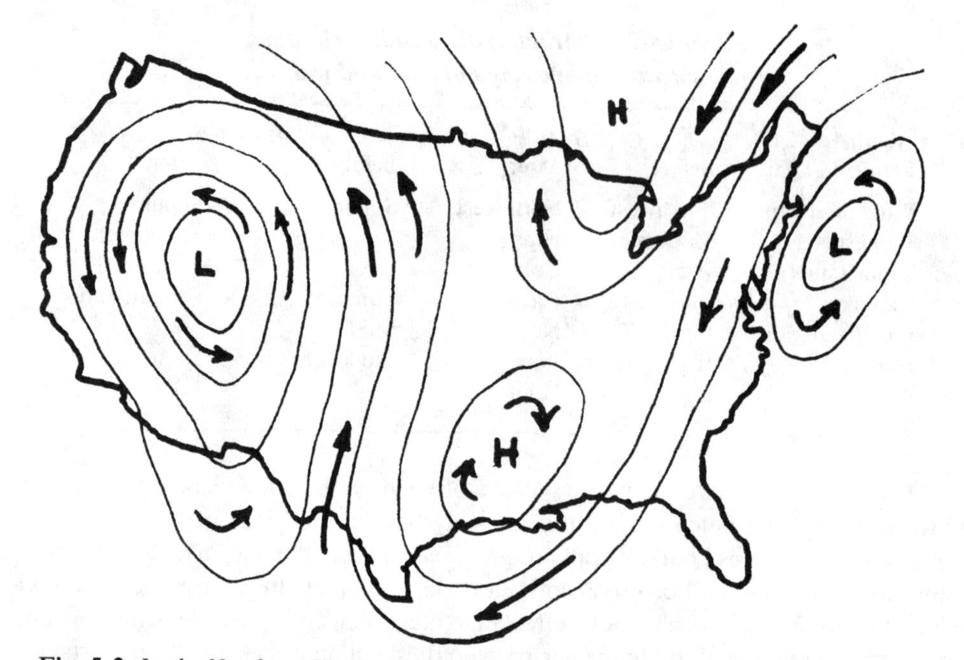

Fig. 5-2. *In the Northern Hemisphere, the general flow of wind well above the surface of the earth is clockwise around a high-pressure area and counter-clockwise around a low-pressure area. On a weather map showing pressure gradients as* isobars, *the wind will be parallel to the isobars.*

The *troposphere* is the layer of air that is characterized by an overall decrease of temperature with increasing altitude. Starting at the surface of the earth, it rises to an average altitude of about 55,000 feet, although this height varies with latitude and seasons. The troposphere slopes from about 20,000 feet over the poles to 65,000 feet over the equator and is lower in the winter than in the summer. The troposphere's average height above the United States is approximately 35,000 feet.

The *tropopause* is a thin layer that marks the boundary between the troposphere and the next layer, the stratosphere. The *stratosphere* is characterized by relatively small changes in temperature and extends to 150,000 feet above the surface of the earth.

The tropopause is not a smooth, continuous boundary around the troposphere, but rather descends in abrupt "steps" from the equator to the poles. Typically, a narrow band of high-speed wind concentrates in these steps, or breaks, in the tropopause. These winds reach speeds in excess of 200 knots and meander through the atmosphere near the tropopause, like wayward streams. These are the *jet streams*. By definition, such winds must be greater than 50 knots to be classified as a jet stream.

Although a single jet stream might circle the globe, it is not constant, nor is there only one. Usually the hoselike jet is broken into segments. A second jet stream is common and sometimes three are over the same land mass.

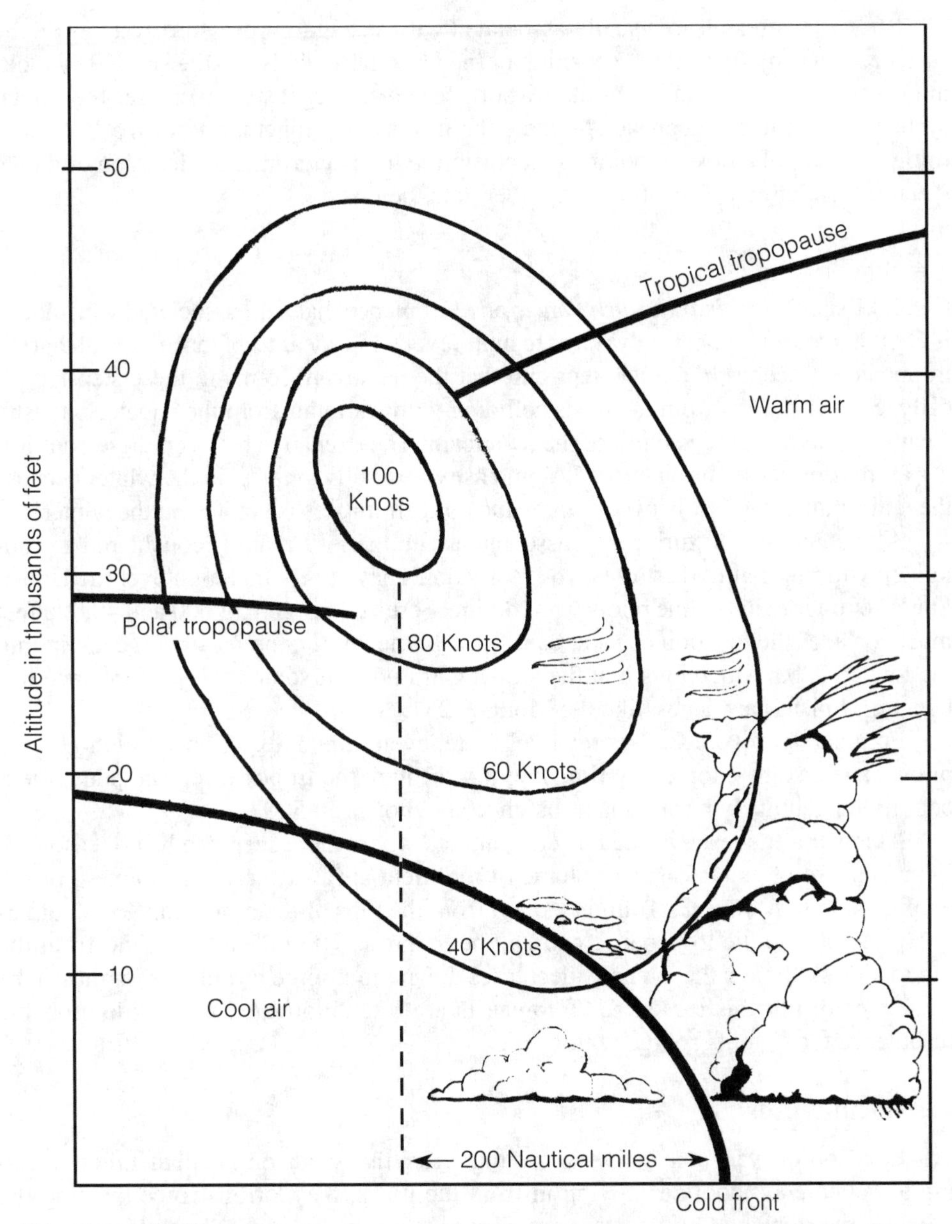

Fig. 5-3. *The* jet stream *is a narrow band of upper-atmosphere wind with speeds greater than 57 mph that is found near the boundary of the troposphere and the stratosphere. Strong jet streams are usually associated with strong low-pressure systems on the surface.*

Strong jet streams are usually associated with strong low-pressure systems on the surface. You can often find the position of the jet stream relative to the surface by looking for an occluded or stalled front on a surface chart. The jet stream crosses the frontal system at the point of occlusion—where the fronts come together to form a Y. Meteorologists routinely look for points of occlusion to help determine the location of the jet stream, correlating the location with other data they have.

CAT

CAT stands for *clear air turbulence*; any turbulence that isn't associated with clouds is CAT, but the term commonly refers to high-level wind shear turbulence. Recall that the tropopause descends in abrupt steps and that the jet stream forms in these steps. Typically, cold air masses from the poles colliding with warm air from the equator intensify weather systems in the vicinity of the jet stream. The exchange between these contrasting air masses causes turbulence. CAT intensity is usually greatest in the winter because the temperature contrast between warm and cold air masses is greatest in the winter.

Although clear air turbulence associated with the jet stream is considered a problem only for high-altitude flights, you really don't have to fly that high over the United States to run into it. Airline pilots who routinely cruise between 30,000 and 40,000 feet must consider the location of jet stream CAT. In the winter, the jet stream can dip into the low 20,000s, which puts it at the cruising altitudes of some high-performance turbocharged piston airplanes, like the Mooney 231.

Fortunately, severe CAT is relatively rare, even though experienced high-altitude pilots expect some chop every time they fly through the tropopause. The odds of encountering genuinely hazardous turbulence are about 1 in 500.

At lower altitudes, CAT can be encountered where there seems to be no reason for it—when strong winds carry a volume of turbulent air away from its source. For example, mountain wave CAT might extend from the tops of the mountains to as high as 5,000 feet above the tropopause and can be as far as 100 miles downwind from the mountains. Although the intensity decreases downwind, some turbulence might still be encountered. For this reason, CAT forecasts are often elongated to include probable turbulence far from its source region.

Local winds

Envision the vast convective forces that drive the winds on the planet: hot air rising at the equator, cool air moving in from the poles, the Coriolis force twisting the winds to the right, and the constant mixing occurring in the middle latitudes. By themselves, these things would make weather complicated enough, but the surface of the earth adds more wild cards to the game. Land and sea are the two biggest players, with different land topographies adding innumerable variations.

Land and sea breezes. The section in chapter 7 regarding overwater flying details how land and sea breezes are formed. To summarize briefly, land warms and cools faster than water. As the land heats up and convection takes place during the day, the air pressure over land becomes lower than over water. A pressure gradient forms, causing the higher pressure air over the water to flow toward the lower pressure air over land. The result is a sea breeze because it comes from the sea.

In the evening, the opposite takes place. The land loses heat rapidly due to radiation, while the water loses heat at a much slower rate. A pressure gradient forms again, this time with higher pressure air over land moving to lower pressure air over water. This is a land breeze.

Mechanical turbulence. Obstructions such as buildings, trees, and rough terrain displace the smooth flow of wind and cause it to move in complex eddies: up, down, and sideways (Fig. 5-4). This mechanical disruption of the ambient wind flow is commonly called mechanical turbulence. The stronger the wind and the rougher the obstructions, the more mechanical turbulence will be created. Strong wind also carries the turbulent eddies downstream, sometimes unbelievably long distances from the source of the turbulence.

Fig. 5-4. *The mechanical disruption of the ambient wind flow by obstructions such as buildings, trees, and rough terrain is called* mechanical turbulence. *Stronger winds and rougher obstructions can create severe mechanical turbulence.*

Be particularly aware of mechanical turbulence at airports. Most of the time, mechanical turbulence caused by buildings is not a problem for aircraft because you fly well above it. But when you operate close to the ground, as you do at an airport, awareness of mechanical turbulence becomes very important. When an airplane is taking off or landing, fluctuations in wind speed and direction can cause dramatic changes in attitude and airspeed. During very gusty conditions, it is smart to maintain a margin of airspeed above normal climb or approach speed to allow for changes in airspeed.

Crosswind conditions require extra vigilance, because these are the times you can expect squirrelly winds coming off nearby hangars or other buildings.

Hangar story

The crosswind at my home airfield wasn't that strong, maybe 10 knots at the most, but it was mainly from the east with only a small southerly component. I crabbed my Taylorcraft down short final to Runway 18, straightened it out with the left wing low just before the flare, and got hit by an updraft.

I have enough problems landing a taildragger in a crosswind, but I knew immediately that the nose-high attitude and ballooning were not due just to my own ineptness. The attitude was decidedly uncomfortable and, although my T-craft does not have a stall warning horn, I knew it wasn't long before the air flowing over the wings would decide to make a break for it. For a moment I considered salvaging the landing, but the thought of falling to the ground from 15 feet in the air and having to pay for the damages myself was just too much to bear. This thought was still lingering in my mind when the unexpected updraft changed to a downdraft, with a few stray eddies thrown in just to make it interesting.

I gave the little 65-hp Continental full throttle—with an engine that's approaching 50 years of age, who knows how much horsepower it gives any more—and pushed the nose down at the same time. The squirrelly wind required some interesting aileron and rudder inputs to keep the heading straight, and I don't imagine anyone on the ground would have been impressed by my flying ability. I didn't care at that point; all I wanted was airspeed. Airspeed is your best friend in a situation like this, and I was looking for such a friend for my sake and the airplane's sake.

Fortunately, the T-craft found some airspeed before the ground found both of us. With the wheels a foot over the grass, we accelerated to a fast climb speed and zoomed upwards for another landing attempt. On downwind, I looked down at the airfield and saw the cause of the unsuspected turbulence—a hangar on the east side of the runway.

The hangar had always been there, of course, but the winds at my home field are usually westerly with a north or south component. As a result, I had flown in crosswinds before, but never one from the east. It didn't take much thought to figure that the wind was curling around the hangar and dumping turbulence over the runway, right where I had flared on the last approach.

It also didn't take much figuring to decide to make a steeper approach and fly over the turbulent air. This I did and, although the crosswind was just as strong, it was steady all the way to touchdown. Before the first landing attempt, I had considered doing some touch-and-gos for practice, but I decided I'd tempted fate enough already. I taxied in and shut down.

Orographic turbulence. A form of mechanical turbulence that concerns the movement of air around mountains is called orographic turbulence. Generally, when the wind moving along the surface encounters the side of a mountain, it flows up the mountain's side, creating strong updrafts. Once over the mountain top, the wind flows downward, creating strong downdrafts. Any irregularities in the shape of the mountain will create areas of turbulence. More information about mountain winds and weather is found in chapter 7.

Storm winds. Storms and wind go hand in hand. The same convective currents that create wind also can create storms; storms, in turn, create their own winds. The more severe the storm, the greater the wind associated with it. Hurricanes approach the global scale and cause changes to the wind over hundreds of square miles. Tornadoes are very localized, but can spin up the wind to velocities that are even greater than hurricanes.

The downward movement of precipitation also creates local winds, even very light precipitation. The winds created by thunderstorms are good examples of winds caused by heavy precipitation. The *plow wind* of a mature thunderstorm is caused mainly by the severe downdraft created when rain starts to fall from the thunderstorm cell.

If you doubt this, go take a shower. Leave the bathroom door open and watch your shower curtain. It moves, doesn't it? The downward movement of the water from the shower nozzle displaces the air directly under it, causing very localized wind currents to form. Now imagine what happens when a heavy downpour from a thunderstorm covers many square miles. Right! The air has to go someplace. The wind created this way might be very localized, but it can be very strong.

Hangar story

A 22-year-old commercial pilot with 273 flight hours took off with a friend from Newton, Kansas, on a practice flight in a Cessna 150. After doing a series of touch-and-gos at a nearby airport, they climbed to 6,000 feet MSL to do some airwork: chandelles, lazy-eights, steep turns, and the like. Although there were a few thunderstorms building in the area, none were big enough or close enough to be of concern.

However, they did notice light rain falling from one cloud that wasn't too far from their position. During one of the maneuvers, they flew into the rain. As the intensity of the rain increased to what the pilot described as "basic, good-natured rain, not really heavy," the airplane began to descend at a noticeable rate.

The pilot checked the vertical speed indicator and was surprised to see it showing 1,500 fpm. He increased to full power (and added carb heat just in case), but the descent continued. Realizing the rain and the downdraft were related, he turned left about 130° to fly out from under the cloud. As soon as they got out of the rain, the descent stopped, but they had lost about 2,500 feet altitude.

So you see that wind is a function of many things, both grand and small. A good pilot is always aware of the wind, whatever its source. She looks for its manifestations in the sky and on the ground. She examines clouds for hints that tell her the strength and direction of the wind and the possibility of turbulence. She looks for signs of wind on the ground, such as the movement of smoke, dust, trees, and other things. (I told my wife that in strong winds, cows stand with their backsides pointing toward the wind, and she laughed at the thought of pilots flying around looking for a herd of cattle to tell them the wind direction before landing. If you're landing at an turf strip out in the boonies, sometimes cows are the only wind indicators you can find!)

Turbulence intensity levels

Reporting the intensity level of turbulence is open to a great deal of interpretation. The most important variable is the size of the aircraft, although type of aircraft and pilot interpretation are factors, too. In most cases, the heavier the aircraft, the better it tolerates turbulence.

According to some sources, turbulence is divided into three classes (with simplified definitions): *light* turbulence, which scares the passengers; *moderate* turbulence, which scares the airline cabin attendants; and *severe* turbulence, which scares the pilots. Table 5-3 lists the accepted turbulence intensity levels and their definitions, as given in the *Airman's Information Manual*.

Table 5-3. Turbulence intensity levels

Light. Momentary, slight erratic changes in altitude and/or attitude (pitch, roll, yaw). Occupants feel slight strain against seat belts or shoulder straps. Unsecured objects might be displaced slightly.

Moderate. Similar to light turbulence but of greater intensity. Changes in altitude and/or attitude occur, but the aircraft remains in positive control at all times. Causes variations in indicated airspeed or rapid bumps or jolts without appreciable changes in aircraft altitude or attitude. Occupants feel definite strains against seat belts or shoulder straps. Unsecured objects are dislodged.

Severe. Large, abrupt changes in altitude and/or attitude. Causes large variations in indicated airspeed. Aircraft might be momentarily out of control. Occupants are forced violently against seat belts or shoulder straps. Unsecured objects are tossed about.

Extreme. The aircraft is violently tossed about and is practically impossible to control. Structural damage is possible.

When evaluating a turbulence report, consider the type of aircraft flown by the pilot who filed the report (Fig. 5-5). Aircraft size and weight are both important, as well as wing loading. *Wing loading* is the total force exerted on a wing in flight and is expressed in pounds per square foot of wing area. A light aircraft with large wings, such as a sailplane, has low wing loading; a heavy aircraft with small wings, such as a jet fighter, has high wing loading. An airplane with higher wing loading handles turbulence in a more stable fashion. Airplanes with light wing loading experience drastic variations in pitch and roll in turbulence.

Cessna Aircraft Company

Fig. 5-5. *When evaluating a turbulence report, consider the source as noted in the report. The pilot of a light airplane will generally report a higher intensity level than the pilot of a heavy airplane.*

Helicopters tend to fly better in turbulence than airplanes of similar gross weights because all the weight of a helicopter is hanging from the main rotor, which is the "rotary wing" of a helicopter. The rotor absorbs the smaller bumps before they are transmitted to the fuselage; in an airplane, the rigid fixed-wings take the bumps and send them directly to the fuselage.

Therefore, when considering a pirep about turbulence, you should also consider the type of aircraft specified in the report. Moderate turbulence reported by the pilot of a Cessna 150 might be little problem for the pilot of a Beech King Air, whereas moderate turbulence reported by a Boeing 747 crew would give the King Air pilot cause for concern.

Turbulence encounters

Light turbulence. In a small aircraft, light turbulence is common. In fact, it is a rare day when you don't encounter any light turbulence during a flight in most parts of the country. Normally, no particular action is required; you just keep on flying and accept the bumps. If the turbulence becomes too uncomfortable or increases to the moderate stage, you might want to try a different altitude. In most cases, this means climbing to a higher altitude, but under some conditions a descent might work, too. A change in flight route could make a difference, as well. For example, if you were flying along a mountain ridge, you'll probably find the air smoother on the upwind side than on the downwind side of the ridge. Or, if you are paralleling a coastline, the air over the water will probably be smoother than the air over the land, particularly during a sunny day. At night, the air over the land might be smoother.

Moderate turbulence. If you fly enough, you'll probably encounter moderate turbulence. The heating that occurs on a warm summer's day in any part of the world is enough to create moderate turbulence for light aircraft. Most pilots get used to moderate turbulence and don't think too much about it. In fact, they might even call it light turbulence. (Light turbulence then becomes the "normal" flight condition, just slightly worse than "smooth as glass.")

Most passengers are different, however. Unless a passenger is a frequent lightplane flier, moderate turbulence is very uncomfortable, often frightening. Not only will inexperienced passengers be more prone to motion sickness, but they'll also begin to worry about their personal safety, as in "Will the wings fall off?" All it takes are a few rapid bumps or a couple of objects being dislodged.

Keep this in mind when you're flying. Moderate turbulence isn't really dangerous. If airspeed fluctuates frequently, you should reduce to maneuvering speed (V_A) to provide protection against overspeeding and overstressing the aircraft and perhaps reduce the severity of the turbulence. Besides this, there isn't much for you to do with respect to the safety of the flight. Your biggest concern will be your passengers and their comfort. If at all possible, you should try to find a smoother altitude or route. Perhaps with some passengers, you might have to land and wait for smoother conditions, such as those found in late evening or early morning.

Severe turbulence. Severe turbulence gets everyone's attention, even the most experienced pilots, particularly when they find their aircraft is momentarily out of control. If moderate turbulence scares passengers, severe turbulence will make them terrified.

Although not infrequent, severe turbulence can usually be avoided by paying attention to weather reports and analyzing conditions. If, for example, the wind is reported to be 40 knots at the mouth of a mountain pass, you can well expect to encounter severe turbulence in the pass, even if it hasn't been reported (probably because no one else has dared fly through the pass yet). If you see a thunderstorm building rapidly 30 miles in front of your track, you can well expect to run into possible severe turbulence if you don't make a course change. And if there's a sigmet for severe turbulence or a convective sigmet, you should pay attention to it.

Weather being as changeable as it is, there will be times when you might encounter severe turbulence when you least expect it. Once, I flew to a fly-in breakfast at an airport located in the mountains of Pennsylvania, fully expecting to encounter at least light turbulence en route because I was heading toward the mountains and into the wind. But except for a few minor bumps, there was nothing.

On the return flight, however, I was about 20 miles from my home airport and had cleared the final ridge by about 3,000 feet when I hit one of the worst downdrafts I've ever felt in a light airplane. My feet came off the floor and my hands flew above my head. An empty soda can, maps, the flight log, and a couple of pencils lifted off the passenger seat. The airplane nosed over and began to pick up speed, but as quick as it had happened, it was over.

I grabbed the yoke, leveled out, and looked around for some sign of the cause of my "momentary loss of control." There was none. (This typical pilot reaction—searching the sky and topography for the cause of turbulence—is similar to that of the nonmechanical motorist who opens the hood of a car when it won't start. Neither the pilot nor the motorist expects to discover something that they can do anything about, but the ritualistic act of looking seems to satisfy a need of some sort.) I had no way of knowing if the turbulence would occur again, so I reduced airspeed and continued toward home. The rest of the flight was smooth.

If you encounter severe turbulence and it doesn't stop after one bump, there are a few other things you can do, besides reducing to maneuvering speed. If your airplane has retractable wheels, extending the landing gear will add drag and help keep the airspeed from building if the turbulence puts you into a dive. (Take care to extend the gear only when the airspeed is below the maximum landing gear extended speed, V_{LE}.) To avoid overstressing the airplane, minimize control deflections. Don't try to chase the airspeed, altimeter, or vertical speed indications. It is better to determine a pitch and power setting that will attain maneuvering speed and adhere to these settings while holding the wings level.

If moderate to severe turbulence is expected on landing and approach, add a few knots to your normal approach speed to protect against occurrences of wind-shear-induced stall or loss of altitude.

Extreme turbulence. It goes without saying that extreme turbulence should be avoided like the plague. You might be forgiven for taking a pilot report of moderate or severe turbulence with a grain of salt if you're familiar with local conditions and flying a medium-sized airplane. But any report of extreme turbulence, even from a lightplane pilot, should be taken very seriously by all pilots. Read the definition again if you doubt this.

If you encounter extreme turbulence, try to get out of it as quickly as possible and use the procedures given for severe turbulence. After landing, the aircraft should be examined by a competent mechanic for any structural damage. If you're a renter, don't be shy about telling the owner that you had been violently tossed about because you don't want to jeopardize the safety of the next pilot or passengers.

Wake turbulence

Wake turbulence is often thought of as a problem only during takeoff and landing, but it can be a hazard during all phases of flight (Fig. 5-6).

The same rules apply; the bigger and faster the airplane, the greater the vortex wake that develops behind it. Depending upon atmospheric conditions, the wake can persist for miles behind an aircraft. The vortex widens and weakens farther away from the initiating source, but a C-5's wake could still cause a few uncomfortable moments for the pilot and passengers of a light airplane from quite a distance.

Most of the transport category airplanes climb above 30,000 feet; therefore, en route wake turbulence is not a big concern if you fly below those altitudes. Be alert for

James E. Hobbs, Lockheed Aircraft Service Corporation

Fig. 5-6. *This extraordinary photograph of a Lockheed HC-130 Hercules clearly shows the large spiral-shaped wake forming behind this 175,000-pound aircraft. The wake is a serious hazard to small aircraft and would even cause pilots of larger airplanes to take notice.*

wake turbulence around terminal areas, however, especially those with Class B and Class C airspace. These terminals are frequented by the big airliners. Pay attention to traffic when overflying or underflying Class B airspace.

If you do fly at the same altitudes as the large airliners, be aware that the wake from the flight at 33,000 feet might find its way down to you at 31,000 feet. Usually it is just a momentary thing, but if it continues longer than normal, it might be worth it to change altitude.

Winds and pireps

A meteorologist at the National Weather Service (NWS) once told me that knowledge of the wind is the most important factor in forecasting weather. He said that if meteorologists knew what the wind was doing at 1,000-foot levels over every part of the planet, weather forecasting could reach nearly 100 percent accuracy.

He wasn't exaggerating, but it is impossible to prove his point. Over land, the NWS obtains wind information from *radiosonde weather balloons* that are released twice a day from some 150 stations in the United States. This information is supplemented from automatic reports that are transmitted from a number of airliners while in flight and new vertical radars, called "wind profilers," which have been installed at locations in some states. Combined together, this gives a "pretty good" general picture of what the winds are doing, but the picture could be improved considerably if more pilots gave more pireps on every flight.

Of course, the information is useless if it is wrong, so don't go making things up. But if you have a loran, GPS, or other system that gives you a good read-out of wind, take the time and provide at least one good pirep on every flight.

IN-FLIGHT ICING

"The ice was here, the ice was there,
The ice was all around. . . ."

Samuel Taylor Coleridge,
The Ancient Mariner

Many texts on weather hazards don't group ice and snow together. There's a good reason for this. Flying in icing conditions is not the same as flying in snow; however, flying in snow can often precede or follow flying in icing conditions. Sometimes it is hard to tell the two apart. Because any frozen precipitation that sticks to an aircraft is cause for concern, both ice and snow are covered here.

Icing types

In-flight icing occurs when an aircraft flies through visible moisture, such as rain or cloud droplets, and the aircraft structure is below the freezing temperature of water (0°C or 32°F). For the IFR pilot, icing is a major concern at some times of the year and a less important concern most of the year, but it can also be a problem for VFR pilots flying in cold rain or drizzle, mist, and snow. Three types of in-flight icing are clear, rime, and mixed (Fig. 5-7). A fourth type of icing of concern to pilots, frost, accumulates when aircraft are on the ground.

Clear ice. Clear ice occurs when the temperature is close to freezing, quite often in cumulus and thunderstorm clouds that are generating large raindrops. Clear ice looks—this is a hard one—clear, as clear as glass. Like glass, it is transparent and has a smooth surface. Freezing rain will also form clear ice. After impact with a cold surface, the liquid portion of the drop flows out over the surface and gradually freezes into a smooth sheet of ice. Subsequent drops freeze on top of the ice and similarly flow over the surface, following the shape of the structure.

Some weather experts tell us clear ice can form from 5°C to –10°C (41°F to 14°F); other experts say it will form from 0°C to –10°C (+32°F to –14°F). Take your pick, believe what you want, and act accordingly. From my personal experience, I've never seen ice accumulate above the freezing point and just 1°C above freezing was always enough to cause it to slough off. On the other hand, I've never flown inside a really big thunderstorm with large super-cooled raindrops either. The safe route is to be alert for icing anytime the temperature is at 5°C or lower and, when the temperature is hovering around the freezing point, become extra alert.

Clear ice can build up quickly, adding weight and disrupting the aerodynamic efficiency of the aircraft. Although it is not as common as rime ice, clear ice that forms due to freezing rain or drizzle can build to hazardous amounts in just a few minutes. A fellow helicopter pilot in Iceland reported a clear ice encounter that caused him to go from using 70 percent power at an airspeed of 100 knots to 100 percent power at an airspeed of 70 knots in order to maintain level flight. Fortunately, he flew out of the icing conditions shortly thereafter because the helicopter would not have been able to maintain level flight much longer.

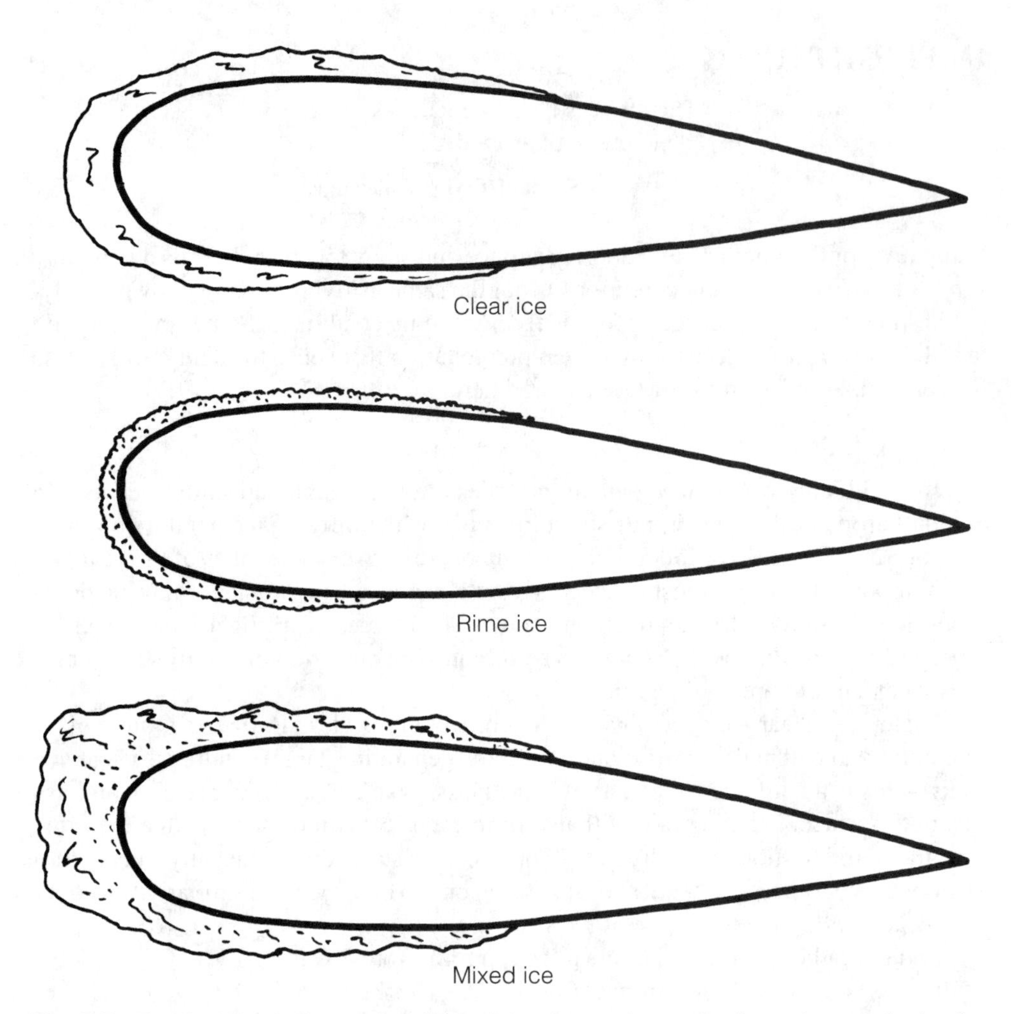

Fig. 5-7. *The three basic forms of airframe icing (top to bottom): clear, rime, and mixed.*

Clear ice changes the shape of the wing and disrupts the airflow over and under the wing. Lift decreases and drag increases. As little as a half-inch of ice on the leading edge of a wing can reduce lift by 50 percent.

Rime ice. Rime ice occurs at lower temperatures than clear ice, usually in the range from –10°C to –20°C (14°F to –4°F) and is common in stratus clouds. Because the water droplets that create rime ice are colder than droplets that form clear ice, they contain less water; therefore, the droplets freeze rapidly and have little, if any, liquid to spread over the aircraft surface. The rapid freezing of the small droplets gives rime ice its characteristic milky-white, almost opaque appearance. It actually looks very much like a coating of light snow or frost, but is much harder and clings tenaciously to the surface. On protuberances, rime ice can grow into fantastic, nonuniform shapes that are quite fascinating to look at.

Rime ice rarely adds much weight to an aircraft, but it can also cause aerodynamic problems for the wings. Although it doesn't spread all over the wing like clear ice can, rime ice can be very difficult to melt off because it forms at such low temperatures. And in spite of what some textbooks say, it is possible to get icing below –20°C (–4°F).

Once in Alaska, I picked up a load of rime ice at –25°C that formed so rapidly I decided to turn around and return to land before exiting the airport traffic area. We were flying VFR, but in a very fine mist consisting of ice crystals. At least one-eighth inch of ice covered the entire aircraft in a matter of minutes. It took three hours in a heated hangar for the ice to melt off. This type of occurrence is rare. In fact, 96 percent of reported icing incidents occur at temperatures above –20°C.

Mixed ice. The third kind of ice is mixed ice, defined as a combination of rime and clear. It results when flying through areas with temperatures around –10°C (14°F) and precipitation with varying droplet sizes. It can also form when icing occurs with wet snow. Mixed ice combines the worst of rime and clear ice and is often associated with stratocumulus and cumulonimbus clouds.

Snow. Everyone knows what snow is. It is that white, fluffy stuff that falls from the sky in winter. What many people don't realize is that you can have snow (as well as icing conditions) in the summer—if you go high enough into the sky. Often what hits us on the ground as rain started as snow higher up. It just melts on the way down.

Hangar story

Some years ago in August, I was headed up the East Coast to Maine. It was bumpy down low, so I decided to go on top. The farther northeast I got, the higher the clouds, so I kept climbing to maintain VFR. I gave up at 11,500 feet and called Boston Center for an instrument clearance. They obliged and I descended to 11,000 feet in cloud and proceeded on course.

About 15 minutes later, I happened to look up from the panel and was shocked to find the windscreen completely iced over. A look out the side window confirmed ice on the wing struts, although it didn't look too bad. The OAT said 29°F.

Boston cleared me down to 9,000 feet and the ice melted off in a matter of minutes. Moral: Yes, you can get ice in the summertime, and you don't have to be in Alaska. (Larry Bothe of Bolingbrook, Illinois)

Generally, the wetter the snow, the more chance you have of it sticking to your aircraft. Dry snow has a hard time sticking to anything. Wet snow has a higher water content than dry snow (because it isn't as cold) and will stick to almost everything. Wet snowflakes are also usually bigger than dry ones. Wet snow will build up on things protruding into the relative air flow, such as pitot tubes, windshields, struts, handles, wheel pants, and the like. It can even stick to the wings. Usually the major problem with an accumulation of wet snow on an airplane is weight (as opposed to changes in wing aerodynamics, which is the main problem with icing) and usually this is more of a problem for slow-movers than fast-movers.

Slow-moving aircraft generally have wings with a wider chord and more protrusions than fast-movers. These wide wings are great for snow build-ups. Because slow-movers aren't built to fly fast, designers are not as concerned about drag-producing protrusions as they are with faster-movers. (Could you imagine wing struts on a corporate jet?) But these protrusions sure are great snow collectors.

The more snow an airplane collects, the heavier it gets. It is possible for an airplane to accumulate so much snow that it can no longer maintain level flight. The airplane might still be getting good lift from the wings, but the added weight of the snow is too much for the engine. This might not be a classic case of icing, but it certainly will create problems for you.

The real danger when flying in snow, however, is that it can easily change into true icing conditions with little warning. All that's needed is a slightly colder temperature. Such a temperature change can happen while flying at the same altitude, or when changing altitude. Now the snow that stuck like wet oatmeal to the wings and sloughed off from time to time suddenly takes on a new character. No longer soft and wet, it freezes like Superglue to the wings and becomes hard, brittle, and crusty.

The result is mixed ice. Mixed ice forms when wet snow or ice crystals form a matrix for clear ice to form around. All icing is a problem, but mixed ice is often the worst. Not only do you get the weight problem, but as clear ice flows over and around the slushy snow that's already there and then hardens, you also end up with an accumulation that can drastically change the aerodynamic efficiency of the wings.

Small, medium, or large?

Like most things in life, icing comes in different sizes. When referring to icing, pilots and meteorologists talk about intensity levels.

Like turbulence intensity levels, icing intensity levels are relative and leave room for much interpretation. Like beauty, icing intensity is in the eye of the beholder. What one pilot considers—and reports in a pirep—as light icing might be considered moderate icing by another pilot. Although ambiguous, this is understandable. Different aircraft accumulate and handle icing differently.

The *Airman's Information Manual* notes: "Different types of aircraft accrete ice at different rates, depending on their airframe shapes and speed differences. For example, temperature rises with airspeed. High airspeeds create surface friction on the airfoils, thus preventing ice accumulation. Consequently, pilots of high-performance jet airplanes might not report icing, but slower-moving aircraft flying through the same area might experience heavy ice accumulations." When checking pilot reports of icing, always consider the type of aircraft that made the report.

Table 5-4 gives the book definitions of icing intensities, straight from AIM.

Note how these intensity levels are defined in relationship to deicing and anti-icing equipment. If your aircraft does not have such equipment, anything greater than trace icing is potentially hazardous; even trace icing can become hazardous if you fly in it longer than an hour. It is just a function of time, because without any way to get rid of or slow the accumulation of ice, the longer you fly in icing conditions the more ice you'll pick up.

Table 5-4. Aircraft icing intensity definitions.

Trace ice. Ice becomes perceptible. Rate of accumulation is slightly greater than the rate of sublimation. (Sublimation occurs when an element changes directly from a solid state to a vapor state; in other words, when ice changes into water vapor without turning into water first). Trace ice is not hazardous, even though deicing/anti-icing equipment is not used, unless encountered for an extended period of time (more than one hour).

Light ice. Rate of accumulation might create a problem if flight is prolonged in this environment (more than one hour). Occasional use of deicing/anti-icing equipment removes/prevents accumulation. It does not present a problem if the deicing/anti-icing equipment is used.

Moderate ice. Rate of accumulation is such that even short encounters become potentially hazardous. Use of deicing/anti-icing equipment, or flight diversion is necessary.

Severe ice. Rate of accumulation is such that deicing/anti-icing equipment fails to control the hazard. Immediate flight diversion is necessary.

If your aircraft is equipped with deice/anti-ice equipment and is certified by the FAA for flight in known icing conditions, it is worth noting the definition of severe icing. You might be fully legal flying in severe icing with your aircraft, but in reality your aircraft might not be able to do it because, by definition, severe icing is so intense that icing equipment is not able to reduce or control the hazard. This is very subjective, too. The icing equipment on one airplane might be fully able to control the hazard while icing equipment on another airplane might not be able to. The first pilot would therefore report "moderate icing" while the second would call it "severe."

You can see that these icing intensity levels leave much room for interpretation, and these are from pilots who actually see the ice building on their aircraft. We haven't even mentioned how difficult it is for meteorologists to forecast any kind of icing, let alone attempt to forecast the icing intensity levels.

How to avoid in-flight icing

To get in-flight structural icing you need two things: visible moisture and a temperature near or below freezing. Don't fly in either one of those, and you avoid icing conditions. That's the easy way. Unfortunately, life isn't always so easy, so the thing to do is to anticipate where icing is going to be, and then avoid it. You can anticipate icing conditions three ways: by examining forecasts, by looking for reports of known icing, and by determining icing conditions on your own.

Forecast icing. Icing is very difficult to forecast accurately, so meteorologists err on the safe side. The National Weather Service uses computer models of the atmosphere that try to compare data obtained from weather balloons and surface observations to conditions that are known to produce icing. Although different models are used for the eastern and western parts of the United States, the conditions used are very general.

In the East, the relative humidity at the surface must be 80 percent or higher and the temperature at the 850 millibar level must be between 0°C and –22°C (32°F and –8°F). With these two conditions present, icing is forecast in the clouds from the freezing level to 18,000 feet.

The Great Lakes and the areas around them are given special attention because the increase in moisture in the atmosphere increases the probability of icing. Over the Great Lakes, most icing occurs below 8,000 feet, so this gets factored into the forecasts.

In the West, the computer model is more complicated. First, the relative humidity at the surface must be 55 percent or greater. With respect to temperature, the computer looks at the difference in height of the 500- and 1000-millibar levels because this is a direct indicator of temperature. Finally, the model looks for areas where air is rising from lower elevations to higher elevations.

Both East and West models tend to overforecast the presence of icing, so the National Weather Service depends greatly on pireps to supplement their forecasts. You should depend on them, too.

Known icing. When a pilot reports icing conditions—forecasted or unforecasted—those conditions become "known." We've already seen how subjective this can be, but a report about an actual icing encounter is always better than a computer model's guess.

The following are my personal rules of thumb for judging other pilots' icing reports:

- Pilot reports of trace icing are usually correct.
- Pilot reports of light and moderate icing are subject to the most interpretation.
- Pilot reports of severe icing should always be taken seriously.
- The less experienced the pilot, the greater the chance the actual icing level will be less serious than reported by the pilot.
- The smaller the aircraft, the greater the chance the actual icing intensity will be less serious than reported by the pilot.
- The less equipped the aircraft is for icing conditions, the greater the chance that the actual icing intensity will be less serious than reported by the pilot.
- Professional pilots flying well-equipped multiengine airplanes for scheduled air services might underreport an icing condition in order to avoid having to delay or cancel a flight.

Besides the interpretation factor, there's also the time element. You should consider how long ago the pirep was filed and what has happened since then. Has there been a frontal passage in the meantime? Was it a warm or cold front? Has the sun risen higher in the sky and warmed things up, or is the air becoming cooler? Has the freezing level changed? Has the storm passed or is it just building? How long did the pilot actually fly in the icing conditions? (Although very relevant, this is usually impossible to determine from a pirep. The most you'll probably find out is if the icing conditions were encountered during climb or descent, neither of which should take very long.)

Determining icing conditions

By using forecasts and reports of known icing, you are taking two important steps in determining whether icing conditions exist. Sometimes it will be crystal clear one way or the other, but often it will be a judgment call—your judgment. The following forecast rules will help interpret the possibility of icing:

- In general, the colder the temperature and wider the dewpoint spread, the less the chance of icing.
- If the temperature is 0°C to –7°C and the dewpoint spread is greater than 2°C, there is an 80 percent probability of no icing.
- If the temperature is from –8°C to –15°C and the dewpoint spread is greater than 3°C, there's an 80 percent chance of no icing.
- If the temperature is below –22°C, there's a 90 percent chance of no icing.
- Cumulus clouds created by solar heating have a 90 percent chance of light icing, and icing will usually occur in the upper portion of the clouds (Fig. 5-8).

Fig. 5-8. *Airframe icing will often occur in the upper portions of cumulus clouds when the outside air temperature is below freezing.*

- Moderate icing can be expected in freezing drizzle, when clouds are in a deep low-pressure area, and in clouds within 100 miles of a cold front.
- Severe icing can be expected in freezing rain.

Realize that you might encounter icing when it has not been forecast or reported by anyone else. On the other hand, icing might be forecast and reported and you might not encounter it. In addition, don't expect the freezing level to be exactly where it is forecast; it can easily differ by plus or minus 500 feet, and even greater variations are not uncommon.

What to do if you encounter ice

If your aircraft is not equipped for icing conditions, get out of the ice. Only bad things can happen if you continue. What you should do depends on the general weather conditions, topography, and your capabilities.

If you are flying IFR in stable air and stratus clouds, climbing to a higher altitude and colder temperatures is usually a good strategy. In these conditions, icing is often confined to a relatively narrow band and if climbing does not get you on top of the cloud deck, it will at least get you into a level where the icing stops. You must know the general weather situation in order to make such a decision. Climb at a faster than normal airspeed to minimize time in the icing conditions.

In cumulus clouds, climbing is riskier unless you know how high the tops are and can get there fast—or stay out of them until you're on top. Icing usually occurs in the upper portions of cumulus clouds and if these clouds are building into thunderstorms, it will probably be impossible to outclimb them. In this case, it is better to descend to warmer air below.

On the other hand, the tops of "fair-weather" cumulus represent the upper boundary of unstable air and a flight at only 500 or 1,000 feet above the tops might be both smooth and ice-free. The problem is that it is often difficult to estimate the tops of clouds accurately from a few miles away. I usually guess wrong by 1,000–2,000 feet lower than they really are, even when I add an additional 1,000–2,000 feet to my original estimate! I'm not sure whether this is due to a visual illusion or the fact that the clouds are growing while I'm trying to fly over them—or both.

The topography of the ground you're flying over is another consideration. Over mountainous areas, a descent to warmer air might be impossible because of high terrain, thereby making a climb your only option. If your aircraft can't climb well above the forecast freezing level, be sure you have a route that will take you down and out of the clouds without fear of running into a mountainside. Expect lifting action to vary the freezing level in ways that are too complex to give general rules for. For example, a glacier or the presence of snow on the north side of a slope can create localized areas of fog and icing conditions, even in the summer.

Over water, you can usually find a layer of above-freezing air below 1,000 feet, unless the outside air temperature is exceptionally cold. The warming effect of the water is usually enough to keep the air immediately above it free of icing conditions. The

North Sea helicopter operator I worked for in Norway lost only one day of flight operations in 20 years because of icing conditions, although there were several days that flights had to be delayed. This included flights as far north as 62° latitude. (The warming effects of the Gulf Stream makes the climate in northwestern Europe much warmer than the climate at similar latitudes in North America.)

A VFR pilot obviously should not climb through clouds to get above icing conditions, but a flight in clear skies on top is preferable to flying underneath in snow or freezing rain. You just have to make sure you will have a hole to descend through at your destination. Knowledge of local terrain and weather patterns is helpful here. For example in Alaska, some mountain range passes are frequently closed by low clouds and fog, but can be easily overflown in clear skies with VFR conditions on both sides of the pass.

However, even instrument-rated pilots can run into trouble if they try to climb through freezing rain to get on top of the clouds, as the following story by Larry Bothe illustrates.

Hangar story

Some years ago, I was in Utica, New York, for Thanksgiving and wanted to fly home to Pennsylvania. The surface temperature was 34°F and we had freezing rain. Points to ponder: field elevation about 900 feet, pilot reports stated clear above 6,000 feet, figure 1,000 feet before the ice starts. That gives me no more than 4,000 feet of ice to climb through, and with my Cessna Skyhawk's average climb rate of 500 feet per minute, I'd be through that in 8 minutes. Of course, the Skyhawk is not approved for flight in known icing conditions, but with the weather at home severe clear, I figured it would be a piece of cake. So off I go.

The flight starts off as planned. I begin picking up a little ice at 2,500 feet. Then the rate of climb drops off alarmingly. But I don't give up easily and stick with it until I'm in real trouble. At 5,400 feet, I find myself in a descent at full power.

I think I'm close to a stall, but I'm not sure where that is because the wing has a different shape and the stall horn is iced over. My wife and son are scared to death, and so am I. I tell Utica I want to return, which they approve, along with an immediate descent and approach clearance. I can't do the instrument approach because I'm still taking on ice at the vectoring altitudes.

I have ground contact (thank God), so I fly to the airport VFR. This is a "hot" approach; I am very afraid I will stall on final. But we make it, and I land without incident.

After I get out of the plane, I walk around aimlessly breaking large sheets of ice off the leading edges. Then I regain my composure, call Utica Approach to thank them for their help, and we get a hotel room for the night.

Moral of the story: Don't mess with freezing rain. It is very difficult to outclimb, particularly in modestly powered airplanes. Ice from freezing rain accumulates unbelievably fast and runs quite far back from the leading edges. It's no fun.

There is one more option for IFR and VFR pilots: the good ol' 180° turn. Don't wait too long to make it, however, because you hope things will get better. Optimism in life is preferable to pessimism, but Murphy's Law applies more often to flying. Unless you are absolutely sure that the conditions will improve ahead—and the only way you can know that is a current accurate report from a reliable source—don't continue, relying on blind faith that things will turn out all right. I don't know what it is about this sort of situation, but when things get so bad that you really need an improvement, something else usually goes wrong.

Deicing and anti-icing equipment

If your aircraft is equipped for icing conditions, use the equipment as it is made to be used. There is an important difference between *deicing* and *anti-icing* equipment:

- Deicing equipment is used to remove ice after it has accumulated.
- Anti-icing equipment is used to keep icing from accumulating in the first place.

The distinction between the two is important because improper use of either type can defeat its function.

For example, inflatable wing boots are *deicing* equipment. You are supposed to allow the ice to build up to a certain thickness (about ¼–⅜ inch) on the uninflated boots, then you switch on the system that directs compressed air into the boots to expand them and break off the ice. After the ice is gone, you switch the system off to allow boot deflation. Then you wait for the ice to build up again and repeat the process as often as is necessary.

If the boots are left inflated in an attempt to use them for anti-icing, it is highly likely the ice will accumulate over the expanded boots. Because the boots are already inflated to their limit, it will not be possible to remove the ice. Deflating the boots after the ice has formed over the expanded boots will only open a small cavity of air between the boots and the layer of ice, rendering the boots useless.

One note of caution with wing boots: Don't expect them to remove all the ice on the wing. Ice will remain on the portions of the wing not protected by boots, and even some portions of the boots might not be completely cleaned.

Conversely, windshield and pitot heaters are *anti-icing* equipment and must be turned on prior to entering icing conditions to prevent ice from forming on the components they protect. If turned on after a layer of ice has formed, the systems might not be able to generate enough heat to melt the accumulated ice on the component.

Don't confuse a windshield defroster with windshield heat or anti-ice. An airplane windshield defroster, like the defroster in your car, is not designed to melt off ice, although both are able to help in certain conditions. A defroster's job is to keep condensation off the inside of the windshield. In moderate to heavy icing conditions, a defroster might be just about worthless at keeping the hard stuff away. Even a small amount of trace icing or snow is enough to totally obscure the view forward, and if you've ever made a landing by looking out the side window, you'll know how tough that can be.

A windshield anti-ice system usually involves electrical heating of a thin metallic film that is embedded in the windshield. It works well, but is better at anti-icing than deicing. In other words, when it is on and warmed up, it will keep ice from accumulating on the windshield, but if a layer of ice has already formed on the windshield, the heat might not be sufficient to melt it off completely. If icing conditions are anticipated, the windshield anti-ice system should be turned on before they are encountered; at the very latest, it should be turned on at the first signs of ice on the windshield.

Remember to turn on the pitot heat well before entering icing conditions, too. Check your aircraft manual for restrictions concerning pitot heat; there probably aren't very many. Many pilots seem reluctant to use pitot heat for some reason. When I flew in Norway, we used pitot heat all the time and never had any ill effects. The pitot heads are very sensitive to icing conditions and, in moist conditions, ice can form inside pitot tubes without ice forming on the outside of the aircraft. Iced-up pitot tubes have been the cause of a number of unfortunate accidents when the pilots didn't realize what had happened.

The airspeed indicator is the most significantly affected pitot-static instrument in icing conditions. If the pitot tube becomes blocked, the airspeed indicator no longer receives ram-air pressure. Because the indicator still gets static pressure, it performs like an altimeter, showing an increase in altitude as an increase in airspeed and a decrease in altitude as a decrease in airspeed.

Although rare, static ports that complement the pitot tube can become blocked by snow or ice. When this happens, the airspeed indicator, altimeter, and vertical speed indicator will all "freeze."

More icing precautions

- When airframe icing changes the shape of the wings, you have no clue how your aircraft will fly. Figure that the stall speed is probably higher than normal and avoid abrupt maneuvers. Fly the landing approach with power and faster than the normal approach speed for the same reason.
- Icing on the tail can cause a serious degradation of the tail plane to generate the negative lift needed to keep the nose up. It is suspected that the December 1993 fatal crash of a Jetstream 31 commuter in Minnesota, which occurred when freezing rain was observed on the surface, might have been at least partially caused by the accumulation of severe ice on the airplane's tail.
- Be cautious with flap extension. Many operating manuals warn against the use of flaps because elevator effectiveness might be lost.

Carburetor icing

As air flows past the venturi of a carburetor, the decrease in pressure and the cooling effect of fuel vaporization causes the temperature of the air in the carburetor to drop as much as 70°F less than the temperature of the incoming air. If the air is moist, the moisture might condense and accumulate in the carburetor intake duct as frost or ice. Any restriction to the air flow through the carburetor causes a reduction in power; in extreme cases, ice can totally block the carburetor throat, effectively blocking the inflow of air, and causing the engine to stop (Fig. 5-9).

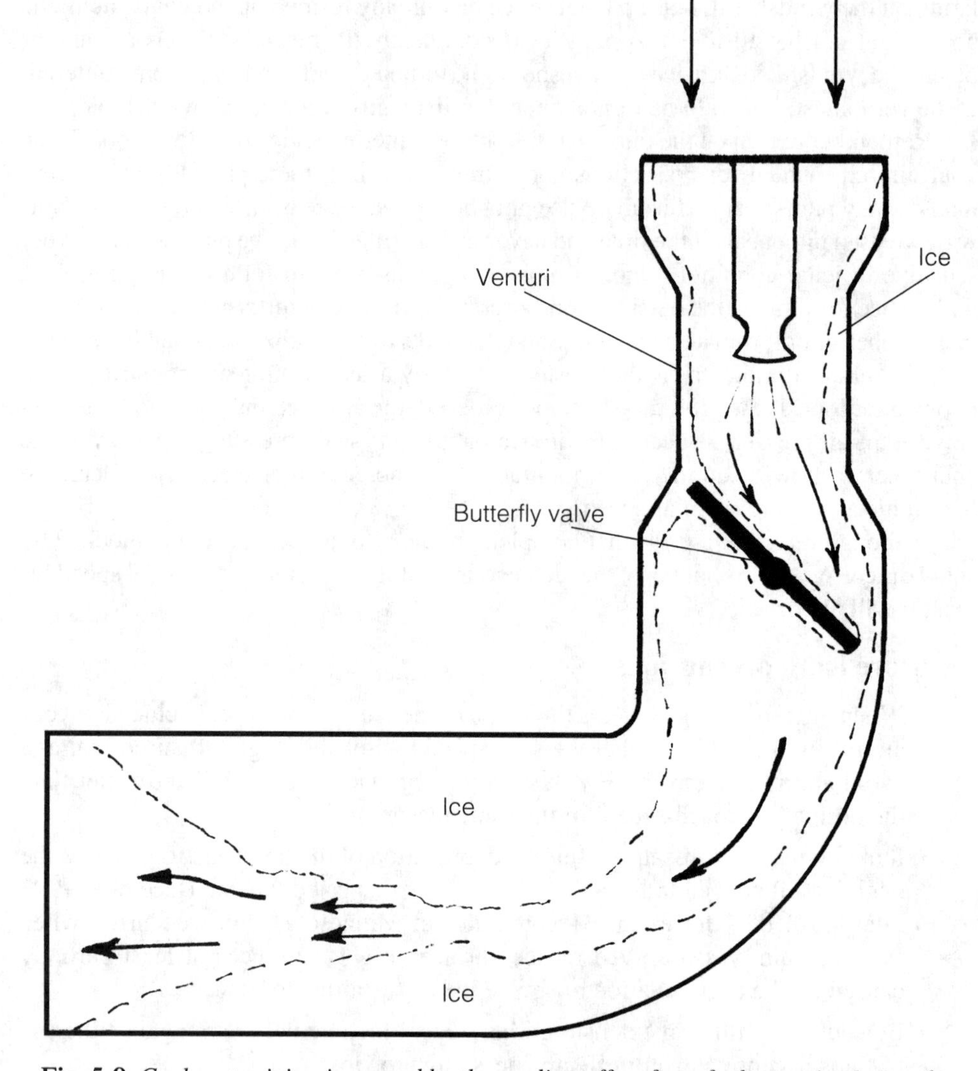

Fig. 5-9. *Carburetor icing is caused by the cooling effect from fuel vaporization and a decrease in pressure as air flows past the venturi in the carburetor. If the air is moist, the moisture condenses and freezes in the intake duct, butterfly valve, and other areas, restricting the air flow to the engine. Segmented lines indicate typical areas where ice accumulates inside the carburetor.*

Carb ice can occur even when the temperature is as high as 90°F and there is no visible moisture in the air, but it is more common when the temperature is below 60°F and the air is humid. It is also more likely to occur at low power settings. The indications of carburetor ice are decreased engine RPM with a fixed-pitch propeller, or a decrease in manifold pressure with a constant-speed propeller.

Carburetor heat prevents and eliminates carburetor icing, making it a dual anti-icing and deicing system. Of course, it is usually best to apply carburetor heat before the icing occurs, but it usually works quite well if you apply the heat after the icing has started.

Application of carburetor heat causes an immediate reduction in engine RPM. If ice is present, this reduction will be followed by an increase in RPM in 10–20 seconds as the heat melts the ice away. You should probably continue flying with the carb heat on until the atmospheric conditions become either much warmer or much drier.

If a lot of ice has built up before the heat is applied, which can happen in very cold conditions, the engine might run very rough for a minute or more as the ice melts and is ingested. Be patient and don't panic; let the carb heat do its job and don't remove it too soon in this case. Leave the carb heat on until outside conditions warm up and dry up considerably.

If ice was not present when the carb heat was applied, the power reduction remains constant. If you applied carb heat prior to a descent, you can safely remove the carb heat when you bring the engine back to a normal power setting.

Carburetor heat should be used whenever there is loss of engine RPM or the engine runs rough during long descents and glides as a preventive against icing.

THUNDERSTORMS

> "Finally, when over the Seine, in the suburbs of Paris, a huge black cloud came rolling toward us from the west. Our pilot did not like the looks of it, and it certainly did have a terrifying aspect. We did not have ballast enough to throw out and so go above it, so we decided to go down. . . . Just as we got started, a vivid flash of lightning cut through the inky blackness of the cloud near us, followed almost instantly by a deafening roar of thunder, and the rain and hail came down in torrents."
>
> **Major Henry B. Hersey, describing his first ascent in a balloon.**
> ***The Century Magazine*, March 1908**

It is easy to become very melodramatic about thunderstorms, the most powerful weather phenomena on earth—after hurricanes. For the pilot, however, thunderstorms are a much greater threat than hurricanes because they are encountered much more often. Hurricanes might not always be avoidable on the ground, but it is unlikely that any pilot, except one engaged in weather research, would knowingly or unknowingly take off and fly into a hurricane.

Unfortunately, this is not the case with thunderstorms. Pilots fly into or near thunderstorms every day (Fig. 5-10). Sometimes they do so inadvertently; sometimes they are directed to do so by a controller because there's nowhere else to route the flight. Most of the time, although the ride might be extremely uncomfortable, no damage is done to the aircraft; sometimes the aircraft doesn't make it through the storm.

Fig. 5-10. *The admonition, "Avoid all thunderstorms," is good advice, but with more than 100,000 thunderstorms occurring in the United States every year, it's difficult advice to follow.*

Why thunderstorms are a problem

Thunderstorms are so numerous in certain parts of the country at certain times of the year that they are just about unavoidable. In any given year, more than 16 million thunderstorms occur throughout the world, an average of 45,000 per day, 2,000 at any particular moment. The distribution of thunderstorms is not equal; some areas get a lot more than their fair share, the subtropical oceanic regions getting the most. With respect to land areas, the United States has the rather dubious honor of being one of the world's leaders in thunderstorms, with more than 100,000 thunderstorms occurring within its boundaries each year.

Thunderstorms are unstable and unpredictable. Sure, forecasters can predict general areas of possible thunderstorm activity and radar can tell where thunderstorms are right now, but precise predictions about the intensity of a storm, the place where it will be, and the time it will be there are impossible because thunderstorms are always in a state of change. The thunderstorm that hits Wichita is not going to be the same storm that hits Kansas City, even if it appears to be the same boiling mass of clouds.

All thunderstorms are not created equal. There are monsters that grow higher than 50,000 feet, spawn tornadoes, and wreak destruction in their paths. And others are much more benign, but still hazardous to aircraft, that stay under 10,000 feet, cool off the afternoon heat, and bring welcome rain to parched land.

So the admonition, "Avoid all thunderstorms!," although good advice, is sometimes hard to follow. There might be times that thunderstorms will be unavoidable. Still, it is the best advice there is and for the general aviation pilot, there are very few good reasons not to follow it. Professional pilots flying scheduled routes, charters, or corporate missions sometimes must intentionally penetrate less-than-severe thunderstorms, but they usually do so with good equipment and under the pressures of a schedule. Sometimes even these pilots misjudge the power of a thunderstorm, penetrate a severe storm, and crash. General aviation pilots who are involved in thunderstorm accidents often didn't know what they were getting into.

Experience is valuable when writing about any subject, but is not always necessary. My personal experience with thunderstorms is limited to the smaller varieties. When I flew in the Air Force, strict regulations required that thunderstorms be avoided and I didn't argue. After leaving the Air Force, I flew helicopters over the North Sea for 12 years.

I encountered a wide range of adverse weather, including winds above 60 knots, icing, turbulence, low ceilings and visibilities, fog, snow, heavy rain, all kinds of frontal systems, barometric pressures so low that they were below the scale of the Kollsman window in our barometric altimeters, and thunderstorms. But North Sea thunderstorms are puny when compared to the ones that grow in the United States.

If we couldn't fly around them, we could usually fly under them and sometimes over them. Busting through small to moderate thunderstorms was the last resort, but not much different than flying through an occluded front, which we did often. No offshore flights were delayed or canceled because of thunderstorms in the 12 years I was there.

Since returning to the United States, I've been doing most of my flying in light helicopters and a 1946 Taylorcraft, which is equipped with an engine and hardly much else. As a result, I've become a serious thunderstorm-avoider, even those of the smallest varieties. Learning from one's own experience is great, but learning from the experiences of others—particularly when it comes to adverse flying conditions—is even smarter.

Hangar story

One of the most graphic experiences about a thunderstorm I've come across was written by Lieutenant Colonel William H. Rankin, an Air Force pilot and veteran of both World War II and the Korean War. Rankin had the misfortune of parachuting through a large thunderstorm. The incident occurred during a navigational flight from South Weymouth, Massachusetts, to Beaufort, North Carolina, in an F8U Crusader jet fighter on July 26, 1959. Rankin told his story in an article that appeared in the *Saturday Evening Post* in 1960 and later a book entitled, *The Man Who Rode the Thunder*. It is a first-hand account of what the inside of a thunderstorm is like, from the point of view of a "parcel of air," that theoretical volume of air that meteorologists love to talk about—except that this particular "parcel of air" was a real, live human being.

Prior to the flight, Rankin was told that he might run into thunderstorms, with tops up to 30,000 to 40,000 feet, but he figured he could go up to 50,000 feet to get over them. As he neared Norfolk, Virginia, at about 6 p.m., Rankin saw the large mass of a thunderstorm in front of him, its tops definitely higher than 40,000 feet.

He climbed to 47,000 feet to stay above the storm, but for an undetermined reason, the Corsair's engine seized. A fire warning light illuminated and the aircraft lost power. Rankin decided he'd rather eject than ride the unpowered, burning fighter to the ground. He punched out at 47,000 feet.

"I had never heard of anyone's having ejected at this altitude," he wrote later. "The temperature outside was close to 70 degrees below zero. I had on only a summer-weight flying suit, gloves, helmet and marine field shoes." At first, Rankin felt an intense stinging sensation all over his body, but this was soon replaced by "a blessed numbing." He also experienced a rapid decompression from the change in air pressure. "I could feel my abdomen distending, stretching, until I thought it would burst. My eyes felt as though they were being ripped from their sockets, my head as if it were splitting into several parts, my ears bursting inside, my entire body racked by cramps."

Falling at a rate of 10,000 feet per minute (115 mph), Rankin soon entered the thunderstorm clouds. Surprisingly, the denser air at lower altitude and inside the clouds eased both the cold and the pain of the decompression. Although it was very dark in the cloud, Rankin was able to see the luminous hands on his watch and noted it was 6:05. His parachute, set to deploy at a barometric altitude of 10,000 feet, opened a few seconds later.

Rankin figured it would take ten minutes to reach the ground and began to relax, even though he was still in the clouds. Actually, the barometric pressure in the storm had fooled the parachute's triggering device, and he was probably lower than 10,000 feet, but that was inconsequential. His ride inside the thunderstorm had only begun.

"A massive blast of air jarred me from head to toe. I went soaring up and up and up. Falling again, I saw that I was in an angry ocean of boiling clouds—blacks and grays and whites, spilling over one another, into one another, digesting one another.

"I became a molecule trapped in the thermal pattern of the heat engine, buffeted in all directions—up, down, sideways, clockwise, counterclockwise, over and over. I zoomed straight up, straight down, feeling all the weird sensations of G forces—positive, negative, and zero. I was stretched, slammed, and pounded. I was a bag of flesh and bones crashing into a concrete floor.

"At one point, after I had been shot up like a shell leaving a cannon, I found myself looking down into a long, black tunnel. Sometimes, not wanting to see what was going on, I shut my eyes. This was nature's bedlam, a black cage full of screaming lunatics, beating me with big flat sticks, roaring at me, trying to crush me. All this time it had been raining so torrentially that I thought I would drown in midair. Several times I had held my breath, fearing to inhale quarts of water."

The hellish ride in the storm seemed to last forever. Finally, the turbulence began to diminish. Rankin caught a flash of green earth below him and minutes later smashed into the trunk of the tree. The time was 6:40 p.m.; it had taken over 35 minutes for him to parachute down from 47,000 feet.

Lt. Col. Rankin did not escape without injury. His body was covered in bruises and cuts; numerous ligaments, joints, and muscles were sprained or strained; and he suffered temporary amnesia and loss of equilibrium. During the decompression, his body had swelled so much that the stitched seams of his flight suit left imprints on his skin. Fortunately, Rankin recovered and was soon back in the cockpit again, undoubtedly with a new respect for thunderstorms.

A THUNDERSTORM PRIMER

Two types of thunderstorms are air mass and steady state. Both contain just about every weather hazard known to aviation, although air mass thunderstorms on average tend to be less powerful than steady state thunderstorms. Both types of storms should be afforded plenty of respect from pilots.

Air mass thunderstorms. Occurring at random in unstable air, air mass thunderstorms are often due to surface heating. (These storms occur within an air mass, not, as the name might suggest, when air masses collide.) They are actually "self destructive," because falling precipitation in the thunderstorm induces frictional drag, which retards the updraft within the storm and soon reverses it to a downdraft.

The downward flow of air and rain cools the surface and the lower portion of the storm, cutting off the inflow of water vapor. Without the upward flow of water vapor, the storm loses energy and dies; therefore, these air mass thunderstorms usually last from 20 minutes to 2 hours and produce only moderate gusts and rainfall.

Over land, air mass thunderstorms pop up and reach maximum intensity in the middle and late afternoon. Over water, such thunderstorms are more common after dark as cool air from a land mass (a land breeze) flows over relatively warmer water.

Steady state thunderstorms. Thunderstorms that are usually associated with weather systems (Fig. 5-11) are called steady state storms. They often form into squall lines, produce strong, gusty winds, dump heavy rain and sometimes hail, and last for hours. Fronts, converging winds, and low-pressure troughs at high altitudes create upward motion that spawns these storms. Surface heating helps intensify many such thunderstorms.

Although squall lines often develop ahead of a cold front in moist, unstable air, they can also develop in unstable air far removed from any front. The American Midwest is well known for squall lines. Warm, moist air pushed northward from the Gulf of Mexico is often overridden by westerly cold air from the Rocky Mountains.

Normally there is a layer of relatively stable air between warm and cold air masses—a *temperature inversion* layer in which air temperature increases with height instead of decreasing. This inversion layer prevents the warm, moist air from rising into the colder levels; however, surface heating can create pockets of warm air that

Leo Ainsworth, National Severe Storms Laboratory Employees Association

Fig. 5-11. *All thunderstorms are not created equal. Some, like this monster, grow higher than 50,000 feet, spawn tornadoes, and wreak destruction in their paths. Other storms, although still hazardous to aircraft, stay under 10,000 feet, cool off the afternoon heat, and bring welcomed rain to parched land.*

force through the inversion layer. When these warm bubbles of air reach the cooler air above the inversion layer, they really take off upward and form thunderstorm cells (Fig. 5-12). Such cells often form in a line many miles long.

In a mature steady state storm, precipitation falls outside the updraft (instead of inside as with an air mass storm) and this allows the updraft to continue unabated (Fig. 5-13). Consequently, the updrafts of steady state storms become stronger and last longer than those of air mass storms. The comparatively long life of such thunderstorms (up to several hours) is why they're called "steady state."

Other types. Air mass and steady state T-storms are the basic types, but meteorologists continue to define and classify several thunderstorm combinations. Lt. Col. Rankin's storm, for example, was a *mesoscale convective complex*, a formation of several individual air mass storms that combined into a single, highly organized system. Meteorologists aren't sure why these mesoscale complexes form, but they know such storms are prevalent over the American Midwest in the summer, when 60–100 of them form each year. Mesoscale storms can cover 40,000 square miles and last up to 12 hours.

The largest thunderstorms generate tornadoes, which are concentrated low-pressure whirlwinds that are more powerful than hurricanes although, on the whole, not as

Alan Stats

Fig. 5-12. *A well developed thunderstorm, like this one over Arizona, can rise to over 50,000 feet and contain nearly every adverse flying condition that a pilot can encounter in the atmosphere: turbulence, icing, lightning, hail, heavy rain, snow, and low visibility.*

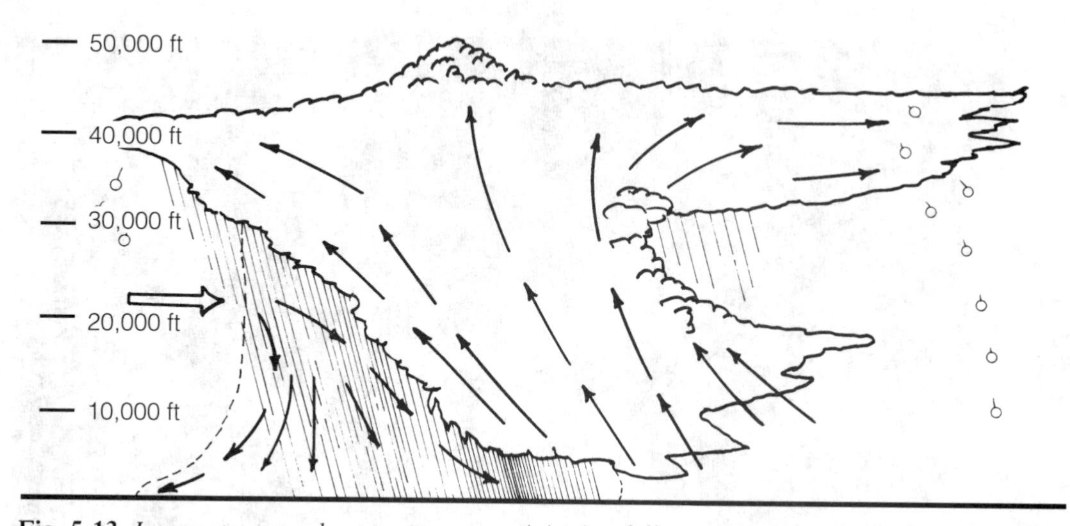

Fig. 5-13. *In a mature steady-state storm, precipitation falls outside the updraft allowing the updraft to continue unabated. The comparatively long life of such thunderstorms is why they're called steady-state. Note hail coming from characteristic anvil top on the right and the gust front (indicated by a dashed line) on the left.*

destructive. The difference is that a hurricane covers a much wider area for a much longer time than a tornado.

Basic life cycle

A thunderstorm cell progresses through three stages: cumulus, mature, and dissipating (Fig. 5-14). These stages are somewhat arbitrary because it is impossible to visually detect when a storm transitions from one stage to another. The changes are often gradual. In addition, what looks like one thunderstorm might actually be a cluster of storms all in different stages of the life cycle.

Cumulus stage. Most cumulus clouds do not grow into thunderstorms, but all thunderstorms start out as cumulus clouds (Fig. 5-15). For weather reporting purposes, a cumulus cloud is not a "thunderstorm" until thunder is heard. Because thunder is the sound given off by the rapid heating and expansion of gases within a lightning channel, the presence of thunder means there is lightning, too.

The main feature of the cumulus stage is the updraft. The updraft starts near the surface and extends to the top of the cloud. Because water vapor begins to condense and form clouds when the air temperature and dew point are equal, you can estimate the height of the bottom of a cumulus cloud by using the surface temperature-dew point spread.

The dew point is the temperature to which air must cool in order for condensation to take place without change in pressure or vapor content. Air becomes fully saturated with water vapor at the dew point, therefore making its relative humidity 100 percent.

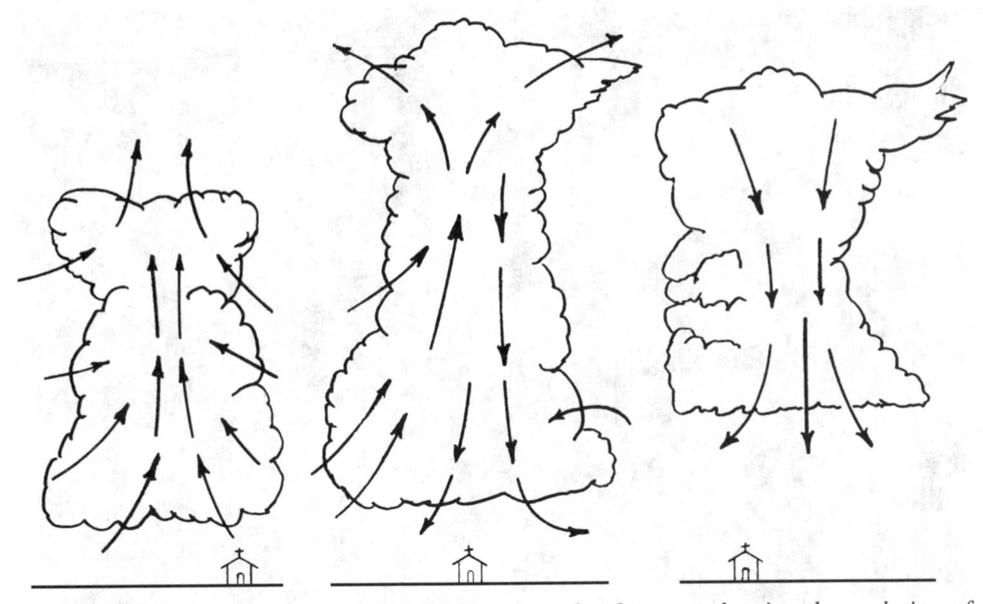

Fig. 5-14. *The three basic stages of thunderstorm development showing the evolution of updrafts and downdrafts (left to right): cumulus, mature, and dissipating.*

Unsaturated air in a convective current cools at a rate of 5.4°F (3.0°C) per 1,000 feet. This rate is the dry adiabatic rate of temperature change and is independent of the temperature of the mass of air through which the vertical movement occurs. The dew point decreases 1.0°F (0.55°C) per 1,000 feet; therefore, temperature and dew point converge at a rate of 4.4°F (2.5°C) per 1,000 feet.

Let's try an example. The surface temperature is 80°F and the dew point is 60°F. Cumulus clouds are forming, indicating the presence of convective currents. The temperature-dew point spread is 20°F (80°F – 60°F). Dividing 20°F by the convergence rate (4.4°F ÷ 1,000 feet) results in 4,545 feet, or about 4,500 feet. The base of the clouds should be about 4,500 feet above the surface.

Note: This method of estimating cloud bases is reliable only with cumulus-type clouds and during the warmer part of the day. Also, cloud tops are much more difficult to predict and outside the realm of us nonmeteorologists.

Early during the cumulus stage, water droplets are small, but they grow as the cloud grows. Cloud droplets collide and merge into larger drops as updrafts carry them higher. The stronger the upward currents, the larger the raindrops and the higher the cloud tops. When updrafts carry the liquid water above the freezing level, it becomes super-cooled. Flying into the tops of cumulus clouds above the freezing level is a fairly sure way of encountering structural icing. Also be aware that the growth rate of a cumulus cloud might exceed 3,000 feet per minute, so it is impossible for most aircraft to out-climb a rapidly building cumulus cloud.

Fig. 5-15. *Cumulus cloud with good vertical development building over a mountain. If conditions continue to be favorable, this cloud will build into a full-fledged thunderstorm.*

Mature stage. When the raindrops grow too heavy to be maintained by the updrafts, they fall. The cold rain drags air with it, creating a cold downdraft coexisting with the updraft. The cell has now reached the mature stage.

When precipitation begins to fall from a cloud, this is your signal that a downdraft has developed and the cell is maturing. Cold rain retards heating and the downdraft remains cooler than the surrounding air. Downdrafts might exceed 2,500 feet per minute. When the down-rushing air reaches the surface, it has to go somewhere and the only place it can go is horizontally along the surface. Strong, gusty surface winds are produced, the temperature drops sharply, and there is a rapid rise in pressure. The surface wind surge is called a "plow wind" and its leading edge is the "first gust."

Meanwhile back in the cell, updrafts reach a maximum, sometimes exceeding 6,000 feet per minute. Updrafts and downdrafts in close proximity create strong vertical wind shear and severe turbulence. Recall Rankin's description: "A massive blast of air jarred me from head to toe. I went soaring up and up and up...I became a molecule trapped in the thermal pattern of the heat engine, buffeted in all directions—up, down, sideways, clockwise, counterclockwise, over and over. I zoomed straight up, straight down, feeling all the weird sensations of G forces—positive, negative, and zero." All thunderstorm hazards reach their greatest intensity during the mature stage.

Dissipating stage. How long a thunderstorm continues in the mature stage depends on numerous factors, but eventually even the most long-lasting "steady state" storms enter the dissipating stage. Strong downdrafts take their toll and cut off the upward inflow of water vapor. When all updrafts are replaced by downdrafts, the storm is dying, although it can still pack a wallop to any aircraft in its vicinity. The dissipating stage is complete when the rain stops and the downdrafts have abated. When all the cells of a thunderstorm have completed this stage, only harmless clouds remain.

Table 5-5. Thunderstorm factoids

Based on a 30-knot speed, the typical thunderstorm will travel about 5 miles in 10 minutes.

The velocity of the surface gusts from a thunderstorm is approximately equal to the sum of the forward speed of the storm plus the velocity of the downdrafts.

The higher the storm, the stronger the downdrafts.

Large raindrops equal large clouds.

The greatest turbulence is found around the edges of a thunderstorm.

In a line or cluster of thunderstorms, new cell generation usually takes place on the southern side while the storms are dissipating on the northern side.

Most thunderstorms in the United States move in a general west-to-east direction. In the Midwest, the storms usually move east-northeast or northeast.

DO'S AND DON'TS OF THUNDERSTORM FLYING

What you do to minimize the risks of flying in and around thunderstorms will depend upon the size of the storm. A small storm can be easily circumvented or even flown under or over. A large storm might require a major detour or a landing and a delay to wait until it has passed. A large group of storms, such as those found in a squall line, definitely requires a landing and a wait of several hours.

Avoiding thunderstorms. To repeat again: "Don't fly into a thunderstorm" is the general rule. Here are some well-proven tips to help you do that.

- Don't land or take off in the face of an approaching thunderstorm. A sudden wind shift or low level turbulence could cause loss of control. Remember, even if you do manage to take off, you might have a problem getting out of the way of the storm and landing back at the departure airport might quickly become impossible. Michael Maya Charles, an airline pilot and contributing editor for *FLYING* magazine, says this about taking off into a thunderstorm, "Truth is, unless you can get well ahead of weather, the absolute safest choice is to wait it out on the ground. In heavy jets, it is best to just set the parking brake, kick back in the seat, and wait it out. There is simply no good reason to go for it, schedules be damned. In smaller airplanes...it is imperative to get information on the approaching weather in time to tie the airplane down securely or get in a hangar."

- Don't attempt to fly under any but the smallest of thunderstorms, even if you can see through to the other side. Turbulence under even medium-size storms can be disastrous. Remember the stages of thunderstorm development: nice, fluffy cumulus clouds have updrafts, but by the time they reach the mature stage, downdrafts have begun. Sometimes you can tell if there are downdrafts by the effects of wind on the surface. Rain coming from any cumulus cloud is a sign of downdrafts.

- Don't fly under virga. Virga is rain that evaporates before reaching the ground. Studies have found that virga often signals the presence of downbursts and microbursts, intense downdrafts of wind that have vertical speeds exceeding 720 feet per minute. (Downbursts affect an area in excess of 2.5 miles; microbursts affect an area with a diameter of less than 2.5 miles.)

- Don't try to circumnavigate thunderstorms covering six-tenths or more of an area, either visually or with airborne radar (Fig. 5-16).

- Don't fly without airborne radar into a cloud mass containing scattered embedded thunderstorms. Scattered thunderstorms not embedded can be visually circumnavigated.

- Don't try to outrun a thunderstorm to your destination. You might beat the storm, but not the first gust or plow wind.

Fig. 5-16. *Don't get suckered into a sucker-hole and don't try to circumnavigate thunderstorms covering six-tenths or more of an area, either visually or with airborne radar.*

- Don't fly anywhere near a thunderstorm unless you are instrument-rated, current, and flying an IFR-equipped aircraft.
- Don't fly at night if thunderstorms are forecast.
- Don't fly into clouds in an area covered by a sigmet or severe thunderstorm watch or warning area.

On the other side of the coin, here are some things you should do:

- Do check the weather for forecasts of thunderstorms.
- Do maintain VFR to observe and avoid buildups.
- Do monitor flight watch (122.0) to learn of current weather conditions
- Do avoid any thunderstorms identified as severe by at least 20 miles. This is especially true under the anvil of a large cumulonimbus.
- Do clear the top of a known or suspected thunderstorm by at least 1,000 feet altitude for each 10 knots of wind speed at the cloud top (Fig. 5-17). Obviously, this exceeds the altitude capability of many aircraft.

Fig. 5-17. *To stay safe when overflying a known or suspected thunderstorm, you need to clear the top by at least 1,000 feet altitude for each 10 knots of wind speed at the level of the top of the cloud.*

- Do regard as severe any thunderstorm with tops 35,000 feet or higher whether the top is visually sighted or determined by radar.
- Do remember that vivid and frequent lightning indicates a severe thunderstorm.
- Do be alert for gusts and wind shear when landing in the vicinity of thunderstorms. Maintain enough airspeed and power during your approach to compensate for variations in airspeed and unexpected sink rates.
- Do make discretion the better part of valor when dealing with thunderstorms. The conclusion, "In retrospect, it is clear the pilot should have turned around earlier, while he was still VMC," is typical of accident reports dealing with thunderstorms.

Before entering. If you can't avoid penetrating a thunderstorm, do the following.

- Tighten your safety belt, put on your shoulder harness, and secure all loose objects.
- Plan your course to take you through the storm in a minimum time and hold that course! Most storms travel west to east, so flying westward, the best bet is prob-

ably to fly straight through. If you're flying eastward, it is best to do a 180° turn before you even enter the storm, but if you don't, try to find the shortest route through the storm. Whatever you do, don't prolong your time in a thunderstorm.

- Establish a penetration altitude below the freezing level or above the level of –15°C.
- Turn on the pitot heat and carburetor heat. Icing can be rapid and cause instantaneous power failure or airspeed fluctuation. Flying in instrument conditions in turbulence is no time to be without reliable airspeed indications.
- Establish the power settings required to maintain turbulence penetration airspeed (usually maneuvering airspeed) recommended in your aircraft manual. Lower airspeed lessens structural stresses.
- Turn up cockpit lighting to the highest intensity to lessen the danger of temporary blindness from lightning. This probably won't make much difference, but it gives you something else to do.
- Disengage the autopilot. It is probably best to "ride out" the turbulence without the autopilot; however, follow your flight manual; some aircraft might penetrate thunderstorms better with one or more autopilot hold modes, such as heading hold, engaged. Altitude hold and airspeed hold are probably two you shouldn't use.
- Tilt the airborne radar antenna up and down occasionally. Tilting it up might detect a hail shaft that will reach a point on your course by the time you do. Tilting down might detect a growing thunderstorm cell that might reach your altitude.
- Notify ATC and ask for a block altitude if severe updrafts and downdrafts make it impossible to maintain your assigned altitude.

When inside. I hope it never happens to you, but if it does, these things might help:

- Keep your eyes on your instruments. Looking outside might increase the danger of temporary blindness from lightning, especially at night.
- Maintain a constant attitude; let the aircraft ride the updrafts and downdrafts.
- Don't change power settings. Maintain the settings for thunderstorm penetration airspeed.
- Don't turn back after you get inside a thunderstorm. A straight course through the storm most likely will get you out of the hazards most quickly. Keep the wings level, and don't let the heading wander. Turning maneuvers will increase stresses on the aircraft.
- Don't blame me, the *Airman's Information Manual*, the FAA, the manufacturer, the National Weather Service, or anyone else if these last-ditch suggestions don't help. You should have avoided flying into the thunderstorm in the first place.

Years ago, *The MATS Flyer*, the magazine of the USAF Military Air Transport Service (the forerunner of the Military Airlift Command), gave this advice: "Thunderstorms, like mothers-in-law, are best respected and avoided. The best advice on thunderstorm flying is DON'T DO IT."

Hangar story

The pilot of a Piper Cherokee 180 had just received his instrument rating and was anxious to try it out, so he decided to make a trip from New Jersey to Georgia with his wife. He called flight service, got a weather briefing, and filed an IFR flight plan.

There was no prediction of thunderstorms in the forecast, but after a few hours of flight, the pilot and his wife found themselves in the middle of a dark cumulonimbus cloud. "There was a lot of turbulence," he recalls, "and lightning on both sides. My wife was really scared, and I was worried, too."

He tried to maintain his assigned altitude of 6,000 feet, but even full throttle couldn't keep the airplane from descending. "I asked ATC for a descent below 4,500 feet because I knew that was below the base of the clouds. They cleared us for a descent, and as soon as we broke out of the clouds, the ride smoothed out," he said.

After the flight, he asked an instructor how he should avoid thunderstorms when they are not forecast. The advice was, "If a cloud looks mean, it probably is, so stay out of it."

"When I thought back on it, I did notice that the cloud looked pretty mean and dark before I flew into it. I just didn't think it would be too bad because thunderstorms weren't in the forecast."

USING RADAR AND LIGHTNING DETECTION EQUIPMENT

Two kinds of radar detection equipment can help a pilot get around thunderstorms: ground-based and airborne. Ground-based radar is available to everyone. Airborne radar is available only to those who can afford it. Both have advantages and disadvantages.

ATC radar

All radars send out electromagnetic pulses and then detect the return of those pulses after they bounce off objects. By measuring the time it takes for signals to return, the radar is able to determine a distance to the objects. Various controls on a radar make it possible for the operator to adjust the intensity and scan of the radar. Sometimes an operator finds it necessary to "tune out" some returns in order to better see the returns he wants to see.

Radar signals are not particularly selective and will bounce off buildings, trees, trucks, ships, hills, mountains, waves, and precipitation. The heavier the precipitation, the better the radar will bounce off. Heavy rain and snow are easy to detect. Light snow and drizzle usually cannot be seen on radar; thus, a radar detects thunderstorms by detecting the moisture in the storms. Turbulence, lightning, winds, updrafts and downdrafts are not detected by radar.

Table 5-6. Thunderstorm precipitation intensity levels

Radar does not detect turbulence, wind, lightning, or hail, but experience has taught meteorologists (and pilots) that greater rainfall intensities in thunderstorms usually indicate greater levels of the above hazards. By correlating the amount of precipitation detected by radar with actual conditions, the following intensity levels were determined as a guide to thunderstorm strength. Each successive level means heavier rainfall than the preceding level as well as an increase in other hazards as noted.

Level 1. Weak storm. Expect light to moderate turbulence with some chance of lightning. Rainfall rate 0.01–0.1 inch per hour.

Level 2. Moderate storm. Expect light to moderate turbulence with some chance of lightning. Rainfall rate 0.1–0.5 inch per hour.

Level 3. Strong storm. Expect moderate to severe turbulence and a good probability of lightning. Rainfall rate 0.5–1.0 inch per hour.

Level 4. Very strong storm. Expect severe turbulence and lightning. Rainfall rate 1.0–2.0 inches per hour.

Level 5. Intense storm. Expect severe turbulence, lightning, organized wind gusts, and probably hail. Rainfall rate 2.0–5.0 inches per hour.

Level 6. Extreme storm. Expect severe turbulence, lightning, extensive wind gusts, and large hail. Rainfall rate over 5.0 inches per hour.

Although controllers are usually willing to advise pilots of adverse weather when they see it on ATC scopes, there are a few problems with this. First of all, ground-based radar used by ATC and the military is designed and used mainly to detect aircraft, not weather. Very simply, the shorter the wavelength and tighter the beam width, the better the presentation of small reflective particles, such as rain. Basic overview: Air route surveillance radar (ARSR) and airport surveillance radar (ASR) use long wavelengths and wide beam widths, making them good at picking up aircraft, but not as good for detecting weather.

Second, detecting weather is not the controllers' primary function. Their main job is to keep aircraft separated, and they might have to spend time with a traffic conflict on another part of the scope while you pick your way through a line of thunderstorms. Providing pilots with guidance around weather is considered an "additional duty," although it is a duty they willingly provide whenever they can.

Third, ground clutter, buildings, and mountains might block the view of a ground-based radar so that it is impossible for the controller to see the adverse weather, even if the radar is adjusted to pick it up. The farther you are from a radar site, the more likely you are to be within an area that is "shadowed" by a building or hill on the radarscope.

And fourth, in order to receive aircraft returns better, the controller often has to adjust the radar so that much of the weather returns are eliminated, or at least tuned down. A severe storm might appear very benign to the controller on the ground.

Controllers can adjust radars to avoid ground clutter caused by trees, buildings, hills, and other objects. The *moving target indicator* (MTI) detects moving targets, which airplanes obviously are. Nonmoving targets (ground clutter) are eliminated from the scope when MTI is switched on.

Unfortunately, MTI also eliminates water droplets when the droplets' horizontal speed relative to the radar is less than 7 knots. This means a controller might not see some areas of light-to-moderate precipitation within 10–20 miles of the antenna. Conceivably, the controller could turn off the MTI to detect the precipitation, but this might be self-defeating because the weather would be indistinguishable from the ground clutter the MTI is designed to mask.

One of the most potentially dangerous limitations of radar with respect to weather detection, and this applies to both ground-based and airborne radar, is *attenuation*. With respect to radar, attenuation refers to the situation in which one weather area closer to the antenna reflects or absorbs so much radar energy that distant weather areas are effectively blocked out and not displayed.

For example, a solid squall line depicted at the far edge of a radarscope could appear to have gaps between a number of thunderstorm cells. These gaps could actually be caused by intense areas of weather that are closer to the radar's antenna. Because these closer-in storms absorb all the radar energy, there's no energy left to "paint" the storms that are in the "radar shadows" of the closer cells. The gaps in the squall line aren't really gaps. Severe thunderstorms might be hiding in the shadows.

It is a judgment call, but I believe ground-based radar operators are more prone to not recognize attenuation problems than pilots who use airborne weather radar frequently. This is not because the pilots are any better at interpreting weather on their scopes; it is because pilots have the advantage of being on a moving platform. Attenuation occurs on airborne radar, but often the "sucker holes" disappear almost as fast as they appear because the position of the aircraft changes. Also, pilots trying to find their own path through a line of thunderstorms are probably more alert to attenuation problems than a controller who is more concerned with the separation aircraft.

Ground-based ATC radar has serious limitations when it comes to thunderstorm avoidance. Yes, you may ask ATC for guidance around severe weather, but neither is it the controller's job nor does she have the best weather detection equipment to do it.

(This situation is improving in the United States as more and more WSR-88D dedicated weather radars are installed and brought into use. The plan is to provide ATC controllers with the information from these radars, as well as from the ATC radars.)

Flight timing

There's another important thing to realize about severe weather and radar controllers. When thunderstorms move in, they eat up airspace, and that means there's less room for aircraft. In all major terminal areas that have fixed arrival and departure routes, a thunderstorm sitting on one or more of those routes means that aircraft must be diverted. During certain times of the day, traffic density is already at its peak in many terminal areas, and one thunderstorm can push the system past saturation, causing numerous delays and reroutings.

A line of steady state storms can cause air traffic problems that take hours, maybe even half a day, to sort out. To stop the flow of traffic into the area, ATC uses gateholds to delay flights before they even take off; if you travel as a passenger on scheduled airlines, you've probably experienced this at one time or another.

Logically, late afternoon is the most common time for thunderstorms and heavy traffic to clash; therefore, try to avoid flying to and from major metropolitan areas in the afternoon during thunderstorm season, if you can.

National Weather Service radar

Dedicated weather radar is designed to find storms, not airplanes. Recall that the National Weather Service is in the process of replacing its older weather radars (WSR-57 and WSR-74 systems) with the newer WSR-88D radars, strategically located to eventually give radar coverage over the whole country. (The number designation refers to the year the radar was developed, and the D stands for Doppler.) The new system will be a vast improvement over the previous system. Because of its Doppler capability, the WSR-88D is able to detect not only precipitation, but wind. As a result, it provides improved detection and characterization of wind shear, turbulence, thunderstorms, hail, storm movement prediction, frontal activity, mesocyclones, and tornadoes.

Weather radar information from the National Weather Service is easily accessible to the pilot. All you have to do is call up flight watch on 122.0 MHz. There's virtually no delay in the information because the weather specialist who talks to you on the radio has the large weather radar screen right in front of her. The specialist can superimpose victor airways on the screen to give up-to-the-minute advice about your route.

Use the service, but don't abuse it. You can call every 10 minutes to get updates on the weather, but don't hog the specialists' time. Do take a moment to make a pilot report; remember, other pilots are probably dealing the same weather problems. (If there aren't any other aircraft out there, then you'd better land, fast.)

Airborne radar

Onboard weather radar is great, but it won't make your airplane a thunderstorm-buster (Figs. 5-18 and 5-19). At best, the radar might make it possible for you to avoid a cell sooner and with a less drastic change of course than if you were flying without radar. The radar will make it possible to avoid areas of heavy precipitation at night. The radar might make it possible to pick your way through a loose line of storms. And the radar will be much better at detecting storms than ATC radar. Don't count on airborne radar to make your aircraft an "all-weather" machine.

Like the National Weather Service radar, airborne radar is designed to look at weather by detecting water droplets in the atmosphere. The larger the droplets, the more radio energy is reflected back to the radar. Because large drops indicate large

Fig. 5-18. *A Beechcraft Bonanza V-35 equipped with a nose-mounted WeatherSync digital weather radar system. Air scoops on each side were added because the antenna displaces the normal engine air filter.*

clouds, radar is good at detecting thunderstorms; however, just when you need radar the most, the scope requires the most interpretation.

Skill and practice are required to adjust a radar to get the best picture of the bad weather ahead. The various controls and features provided for the pilot—gain, intensity, tilt, range, track overlay, and the like—make it possible to adjust the display to obtain different "pictures" of the weather. The newest radars even make it possible to look at a storm vertically, instead of only horizontally; the vertical profile shows the height of the cell and the elevations where precipitation is concentrated.

The operating manual provided with the radar should be studied and understood well enough that you can operate the device without having to refer to the book while flying in solid IMC with moderate turbulence. Classes and seminars are available to teach pilots how to interpret the radar's many nuances, such as how to tell where hail is likely to fall.

Like ground-based radar, airborne radar has limitations. Most radars cannot detect turbulence. They can show areas of heavy rain, and we know that heavy rainfall in a cumulus cloud means turbulence. Some newer radars use computers to examine random motion differences between successive pairs of pulse echoes. Based on Doppler techniques, this *pulse-pair processing* makes it possible to detect areas of moderate-to-severe

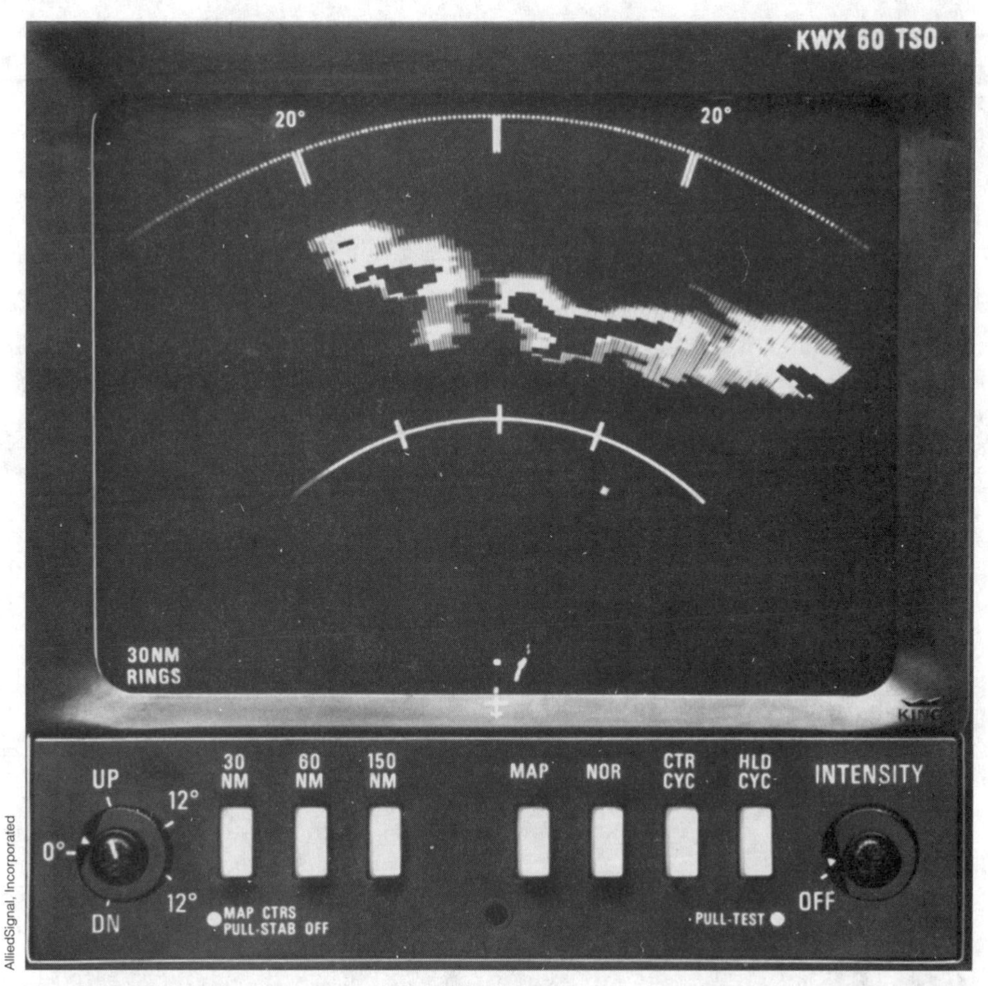

AlliedSignal, Incorporated

Fig. 5-19. *The King KWX 60 weather radar has a 7-inch diagonal screen and automatic tilt stabilization that keeps the antenna at the selected tilt angle during changes in pitch of the airplane. As good as airborne radars are, it takes skill and practice to be able to adjust the radar properly in order to get the best picture of the bad weather ahead.*

turbulence in storms. For the process to work, the presence of measurable precipitation is required, so clear air turbulence cannot be detected. This is an expensive, state-of-the-art weather radar feature; less-expensive radars do not detect any kind of turbulence.

Radars cannot detect areas of icing or light snow. Heavy, wet snow can be detected, but it will be impossible for the radar to tell whether it is snow, or rain, or hail. The radar detects precipitation, period; temperature is undetectable.

Airborne radars are subject to the problems of attenuation, just like ground-based radars. The "hole" you think you see in a line of thunderstorms might be due to the masking effect of a smaller storm between the airplane and the thunderstorms. Newer, more expensive radars include a feature that detects attenuation and displays a warning area behind it. It doesn't tell you if there are thunderstorms in the attenuated area, only that the particular area is not providing you with information, which is better than thinking it is clear.

Finally, radars do not detect lightning. Electrical discharge detection equipment can to that.

Electrical discharge detection equipment

Stormscope is a brand name for the only electrical discharge equipment available in the United States. It was introduced by inventor Paul A. Ryan in 1976, bringing relatively low-cost weather avoidance equipment to general aviation. Ryan sold the patent rights to 3M in 1981, and BFGoodrich FlightSystems bought them in 1991 (Fig. 5-20).

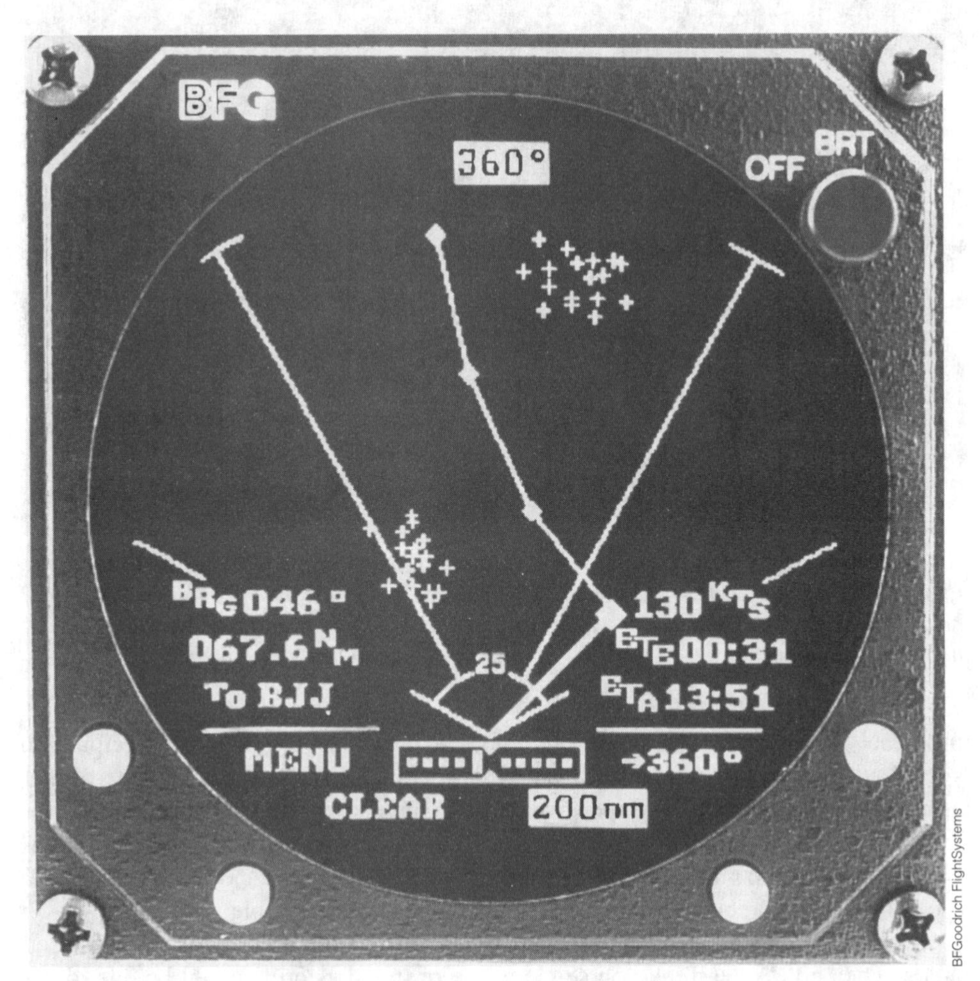

Fig. 5-20. *BFGoodrich FlightSystems Stormscope WX1000 lightning detection system.*

Another device that uses the same principle is the Insight Avionics Strike Finder. Built in Canada, the Strike Finder has run into patent problems with the U.S. courts. BFGoodrich FlightSystems obtained a court-ordered injunction in 1992 that prohibits the Strike Finder from being marketed in the United States because it infringes on the patent that protects Stormscope technology. As this is being written, you still cannot buy a Strike Finder in the United States, though they are available in Canada.

The Stormscope detects electromagnetic disturbances that occur with each lightning strike, which is a very clever way of detecting thunderstorm activity. Recall from the life cycle of a thunderstorm that a cumulus cloud is classified as a thunderstorm when thunder is heard, and that thunder is the sound of lightning. If you avoid the lightning, you avoid the thunderstorm and all the bad stuff that comes with it.

The beauty of the Stormscope is that it will detect young thunderstorms before the rain starts to fall. Stormscopes often pick up electrical discharges before there's enough moisture in a storm for a radar to detect. As a thunderstorm develops and matures, the discharges increase so that a Stormscope also alerts you to the intensity of a storm. Another advantage over radar is that a Stormscope is not subject to attenuation or ground clutter because its receiver does not depend on return signals like a radar.

A Stormscope does have limitations. It detects neither precipitation nor turbulence directly; however, electrical activity and turbulence often occur together in a thunderstorm. By detecting electromagnetic disturbances, a Stormscope gives you a good idea where some turbulence is.

Because a Stormscope does not detect precipitation, when a thunderstorm reaches the dissipating stage and the electrical activity has stopped, there will be no indication of the storm on the Stormscope. Often there is still intense rainfall from a storm at this stage, probably with downdrafts and turbulence. A radar would "paint" the storm because of the precipitation, but the only way you would know of the storm's presence from a Stormscope would be if you had been keeping careful track of the appearance and disappearance of the electrical discharge points.

Radar and Stormscope in the cockpit make a great team for helping a pilot avoid thunderstorms. Understanding their strengths and weaknesses allows a pilot to make the sum of the two even greater than their individual parts.

Nevertheless, remember what you have is great thunderstorm *avoidance* equipment, which will help keep you out of harm's way only if you interpret them properly. If you really want to penetrate thunderstorms on a regular basis, get a job as a weather research pilot.

LIGHTNING

> "It is estimated that somewhere around 1,800 thunderstorms are operating over the earth's surface at any given moment and that lightning strokes strike the earth about 100 times per second."
>
> **Charles R. Roth, *The SkyObserver's Guidebook***

Because a cumulonimbus cloud officially becomes a thunderstorm when thunder is heard, and because thunder is the sound produced by the explosive expansion of air heated by a lightning stroke, thunderstorms and lightning go together like bacon and eggs—or, if you are watching your cholesterol, like oat bran and skim milk. Lightning is an effect of electrification within a thunderstorm.

A tale of two theories

As a thunderstorm develops, interactions of charged particles produce intense electrical fields within the cloud. Scientists have yet to entirely determine how the currents in cumulonimbus clouds produce electrical charges and then separate the positive from the negative in different regions of the cloud. One theory, the "precipitation" theory, suggests that it has to do with the development of ice crystals and snow pellets.

Precipitation theory. It is accepted as fact that moisture in the air condenses as convective currents carry it upward, and that if lifted above the freezing level, the moisture forms ice and snow. According to the precipitation theory of lightning development, continued convective action and the opposing force of gravity causes these particles of snow and ice to move around and collide. When snow and ice particles bump into each other, they exchange electrons and become electrically charged. Before lightning can occur, however, the positively charged particles must separate from the negatively charged particles. Because the larger snow pellets are heavier and tend to be negatively charged, they fall toward the bottom of the cloud. The lighter, positively charged ice crystals are held higher by the convective currents, congregating in the upper part of the clouds.

Convection theory. The other theory, sometimes called the *convection model*, proposes that updrafts carry positive charges that are normally found near the ground upward, while downdrafts simultaneously carry negative charges found in the upper air downward.

Both theories explain some observations, but neither accounts for everything that's known about lightning. Some scientists believe a combination of both models might be the solution. Whatever the case, it is known that a large positive charge is usually concentrated in the frozen upper layers of the cloud, and a large negative charge, along with a smaller positive area, is found in the lower portions. With the positive and negative charges separated, an electrical field is generated. When this field becomes strong enough, electricity flows between the positively and negatively charged parts of the cloud.

Cloud-to-cloud/cloud-to-ground

The first lightning strikes occur within a developing thunderstorm. They can also go from one cloud to another, or one storm cell to another, covering distances as great as 10 miles. In fact, 90 percent of the lightning activity of thunderstorm clouds occurs either inside the clouds or between one cloud and another cloud.

The other 10 percent of lightning is cloud-to-ground. Earth is normally negatively charged with respect to the atmosphere, but as the thunderstorm passes over the ground, the negative charge in the base of the cloud induces a positive charge on the ground below and for several miles around the storm. The ground charge follows the storm like an electrical shadow, growing stronger as the negative cloud charge increases. The attraction between positive and negative charges makes the positive ground current flow up buildings, trees, and other elevated objects in an effort to establish a current flow. But air, which is a good insulator and, therefore, a poor conductor of electricity, insulates the cloud and ground charges, preventing a flow of current until huge electrical charges are built up.

Cloud-to-ground lightning occurs when the difference between the positive and negative charges—the electrical current—becomes great enough to overcome the resistance of the insulating air and forces a conductive path for current to flow between the two charges. This usually happens some 5–30 minutes after the first lightning activity inside the cloud occurs. Several rapid leader strokes of negative charge initiate a path in the sky and are instantly followed by a ground-to-cloud streamer carrying a positive charge. The flows continue until the electrical field is reduced below the level needed to overcome the insulating effect of the air. When a thunderstorm begins to dissipate, the in-cloud lightning decreases, and more cloud-to-ground strikes occur.

Electrical potential in lightning strikes can be as high as 100 million volts, which is quite enough to vaporize that oat bran and skim milk breakfast, and perhaps you as well. The average thunderstorm dissipates about a million kilowatts of electrical energy. Lightning strokes proceed from cloud to cloud, cloud to ground or, where high structures are involved, from ground to cloud.

The enormous electrical current in a lightning stroke causes the temperature of the air to rise to millions of degrees along the lightning path. This results in an explosive expansion of air that creates a very temporary vacuum. Nature abhors a vacuum, and cold air rushes in to replace the hot air that was driven away. The movement generates vibrations that are heard as the crack or rumble, depending upon one's position in relation to the lightning strike.

Although nearly all lightning is generated by cumulonimbus clouds, there are rare cases of lightning occurring in clear, cloudless skies. The cause for this kind of lightning is unknown, but it is suspected that it occurs when a thermal layer overrides another thermal layer of a different density, and the molecular friction generates an electrical potential.

It is estimated that lightning strikes the Earth about 100 times per second. In the United States, the death toll from lightning averages 80 people per year, higher than both tornadoes and hurricanes. (Floods cause the most weather-related fatalities per year at 146.) In addition, more than 250 people are injured each year in lightning-related accidents.

St. Elmo's fire

Another form of electrical discharge is often called St. Elmo's fire. Whereas lightning is a rapid discharge of electricity, St. Elmo's fire is a slow discharge of natural electrical currents. It appears as a blue glow of flickering light on the tips of projections in the air.

Also called *brush discharge* or *corona*, St. Elmo's fire gets its name from San't Ermo of Erasmus, the patron saint of Mediterranean sailors. "St. Elmo" is a corruption of the name to Italian. Sailors apparently regarded the "blue fire" that formed on the masts of their ships as a visible sign of the saint's guardianship over them.

The collision of solid particles in snowstorms and sandstorms generates electrical currents between the sky and ground and causes the phenomenon to occur. Brush discharges can extend into the atmosphere from tall projections, such as antennas, ship masts, or mountain peaks. Sometimes they even extend from animals or people.

On aircraft, St. Elmo's fire is often seen on the tips of the propellers, pitot tubes, antennas, and wing tips. In one reported incident, the pilots saw the corona growing from the radome on their Boeing 727's nose, nearly to the length of a telephone pole. After a muffled explosion and a blinding flash, the flight attendant reported seeing a huge, spinning ball of fire rolling down the aisle at about a hundred miles per hour. Ernest Gann in *A Hostage to Fortune* described a display of St. Elmo's fire while flying over the Atlantic: "All of the C-87 we can see—the wings, engine cowlings, propellers, and bulbous snout ahead—is glowing with an iridescent light that builds and fades and is sometimes so brilliant it seems we are flying inside a second, much larger aircraft."

In a helicopter, St. Elmo's fire sometimes occurs on the tips of the rotor blades. It might occur when flying in dry snow, ice crystals, and in the vicinity of thunderstorms.

Far from being a sign of guardianship, St. Elmo's fire often precedes a static discharge that can be damaging to an aircraft.

Static discharges and aircraft

Although lightning doesn't really *strike* an aircraft, it can and does pass through aircraft. Most reported lightning strikes in aircraft are actually static discharges. From a point of view inside the aircraft, a lightning strike and a static discharge will look much the same. Although the exact causes are different, the effects are much the same, too.

A *static discharge* occurs when an electrical charge builds up between an aircraft and the surrounding cloud. When the charge becomes great enough, a discharge occurs (Fig. 5-21).

Often, but not always, a static discharge will be preceded by St. Elmo's fire. This is one signal that a charge is building, and you should be alert for a discharge. Flight in dry snow and in clouds below the freezing point has proven conducive to the development of an electrical charge between the aircraft and the surrounding elements. Most discharges occur between 10°C and –10°C, and some 30 percent occur right at the freezing level; therefore, if St. Elmo's fire begins while you are flying in these conditions, an immediate change of altitude or course to take you out of these conditions might be warranted.

National Severe Storms Laboratory Employees Association

Fig. 5-21. *Only 10 percent of lightning is cloud to ground. It occurs when the difference between the positive and negative charges becomes great enough to overcome the resistance of the insulating air, which forces a conductive path for current to flow between the two charges.*

The damage caused by a static discharge is usually limited to a few small burn holes in the skin of the aircraft. Likely places are wingtips, ailerons, elevators, rudder, and the nose. Occupants inside an aircraft are usually unharmed because the electrical current travels on the outer surface of the fuselage, and there is little likelihood of sparks or current taking a shortcut through the interior of the aircraft. An aircraft, like an automobile, functions like a *Faraday cage* to protect occupants from the discharge (Fig. 5-22). (Physicist and chemist Michael Faraday designed the protective device.)

However, damage to aircraft is sometimes more than simple burn holes, which are bad enough in themselves. Protruding antennas have been damaged, some completely destroyed. Before lightning-strike provisions were developed for nonmetallic radomes, several airplane radomes were destroyed by static discharges. Compasses and radios can also be rendered useless. Aircraft built from composite materials must incorporate some method of protection from lightning strikes before being certified by the FAA for instrument flight.

Fig. 5-22. *The Glasair III-LP is the world's first kit-built, composite general aviation airplane with verified lightning protection. Here it is undergoing a 1.4-million-volt strike test in the simulated lightning laboratory of Lightning Technologies.*

Why should you avoid an electrical discharge if there is little danger of injury because of the Faraday-cage principle? The hangar story below answers the question.

Hangar story

Two Sikorsky S-61 helicopters were returning from an offshore oil field in the North Sea. One flew at 1,500 feet over the water and the other was at 2,500 feet. It was a clear, moonless winter night with the freezing level at 1,000 feet. About 40 miles from land, a small cloud appeared on the weather radar, right on the flight track. The radar return was very weak, indicating that the cloud didn't have much moisture, which was logical because it was above the freezing level and composed of ice crystals.

The first helicopter flew through the cloud at 1,500 feet, the highly-experienced instrument-rated crew thinking nothing of it. A few minutes later, the second helicopter, with a crew just as experienced, entered the outer fringes of the cloud at 2,500 feet.

The first thing the pilot and copilot in the second helicopter noticed was St. Elmo's fire on the tip of the HF whip antenna that extended forward from the nose of the helicopter. Then they noticed St. Elmo's fire on the pitot tubes and temperature probe, but this was not unusual and neither thought much about it, except to comment that it was there. There was an slight increase in static on the VHF radio and a large increase on the HF radio, but they weren't using the HF radio, so this presented no problem.

One of the pilots turned on the landing light, which immediately reflected off the ice crystals in the cloud. This was fun to watch for a little while, but because it becomes mesmerizing and distracting if you watch it too long, the pilot switched the light off. The pilots also checked for ice forming on the windshields and sponsons, but there was none, which gave them no pressing reason to change altitude.

A minute or so later, St. Elmo's fire formed on the tip-path plane of the rotor blades. This was something neither pilot had ever seen before. Flying offshore, particularly at night, can become somewhat monotonous, so the pilots were enjoying the diversion. They noted that the blue glow on the HF antenna and pitot tubes grew longer. More glowing was forming on the tips of the rotor blades. Both were looking outside the cockpit when the electrical discharge occurred.

Without any other warning, a brilliant flash of pure white light surrounded the helicopter. It was over in an instant, but it was enough to nearly blind both pilots. The captain remembers looking toward the instrument panel and only being able to distinguish the glow of the bright yellow caution and advisory lights. He was completely unable to read any instruments.

The copilot was designated the flying pilot on the return leg. Fortunately, he did nothing to the controls and simply let the helicopter fly along on its own. Just as fortunately, there was no apparent disturbance to the automatic flight control system or engines.

The white flash had been soundless (although the noise of the helicopter could have masked any sound that accompanied it), but immediately after the flash the pilots heard an increase of static in their headsets. After confirming with each other that they were both all right, the pilots agreed not to touch any switches or make any control changes until their sight returned. It did, slowly, and the captain saw there were no caution or advisory lights on the panel that hadn't been there before the flash. Gradually, the two pilots could make out the flight and navigation instruments and noted nothing unusual, except that the ADF needle was swinging back and forth, and there were a lot more returns on the weather radar.

Being suspicious of damage to the helicopter and possible incorrect indications from the navigation and flight instruments, the captain declared an emergency and received radar vectors and priority handling toward the company's heliport. The ADF was obviously useless for navigation. Although the VORs and compasses gave no indication of failure, the pilots didn't trust them. Because there was little other air traffic, and they soon had good visual contact with the coastline after exiting the cloud, declaring an emergency didn't save any time, but it did let the authorities know there was the possibility that a more serious problem might develop. Fortunately, nothing else happened, and the flight ended safely.

Inspection of the helicopter on the ground revealed small burn holes at the tips of each of the five main rotor blades and the tips of each of the six tail rotor blades. As an early warning of possible damage, the main rotor blades on the S-61 are filled with nitrogen gas under pressure; if the gas escapes due to a crack or hole in a blade, a sensor notes the decrease in pressure and causes a warning light to illuminate in the cockpit. No blade warning had illuminated because the burn holes occurred in the very tips of the main rotor blades that are solid steel; however, maintenance personnel prudently determined that it was unsafe to use rotor blades with even such "minor" damage, and removed and replaced all the main rotor and tail rotor blades on the helicopter.

St. Elmo's fire is not just something pretty to look at. It can be a warning of the buildup of an electrical field around the aircraft. Such a buildup can lead to a static discharge. The first helicopter flew into the same cloud about five minutes before the second helicopter. It came out the other end with absolutely no damage. The second helicopter had to have all its rotor blades replaced, a very costly repair. The only difference between the two flights was 1,000 feet of altitude and a static discharge.

A static discharge or lightning strike is obviously something pilots must take seriously.

If you can't avoid lightning

Thunderstorms are in the area, lightning is all around you (Fig. 5-23), and St. Elmo is doing a number on the tips of the wings. What should you do now?

- Turn up the cockpit lights. Assuming it is night, make the lighting in the cockpit as bright as possible. You want your eyes to be adjusted to the brightest possible light in case a flash temporarily blinds you. This is not the time to try to save your night vision by keeping the lights low. In the daytime, it can get dark in the vicinity of thunderstorms, but you probably won't have the same problem with blindness as you do at the night.
- Don't look out the window to admire the glow of St. Elmo. I know it is tempting because I've done it, but you're setting yourself up for a big flash. Keep your eyes inside the cockpit. If you do note a flash, try to shield or close your eyes for a second. The flash is instantaneous, and you might just avoid being blinded. A few aircraft accidents are suspected to be partially caused by blindness of the pilots.
- Keep your hands on the controls. Depending on turbulence, the autopilot may be on or off, but keep your hands on the controls just in case a discharge causes problems. Try to maintain constant heading, altitude, and airspeed.

After the strike is over

Okay, you've done the above and seen the flash. You are slightly blinded, but your vision returns quickly. The aircraft is still flying, but you hear a great deal of static in the headset.

- Check your flight instruments. Your first task is to keep the aircraft flying. Check that the airspeed and altitude readouts are the same as they were before the strike.
- Check your heading. Compare the heading indicator to the compass. Either one could be giving erroneous readings. Do you have the same heading you had before the strike?

Fig. 5-23. *Multiple lightning strikes in a short period of time at night give your eyes very little time to adjust.*

- Test radio and electrical equipment. You'll probably hear an increase in static from the communication and navigation radios, but this will eventually fade as you fly away from the area. If it doesn't, the unit might be damaged. Be suspicious of navigation equipment. Cross-check it against the compass and other equipment.
- Check for tripped circuit breakers. A popped circuit breaker is an indication of an over-voltage on the circuit. The electrical component might be damaged; if you're lucky, the circuit breaker will have done its job and saved the rest of the circuit. Push the breaker in. If it stays in, try the other components on the circuit to check their operation. If the circuit breaker won't stay in, or stays in for a few seconds and pops out again, consider that circuit and its components inoperable, and don't try to use them again.
- Make an entry in the maintenance log, and have the aircraft inspected by a qualified mechanic after landing. It is often difficult to tell the extent of the damage. What looks like a small burn hole to a pilot could be determined to be serious structural damage to a mechanic.

INADVERTENT VFR INTO IMC

"He was halfway through the pass when the ceiling 'came tumbling down' again. 'I found myself flying blind in a pass so narrow that I couldn't turn around. I knew I was in bad trouble.'"

Beth Day, *Glacier Pilot, the Story of Bob Reeve*

Inadvertent visual flight into instrument conditions is not a weather hazard, per se, because clouds are not hazardous, they are just condensed water vapor. Even low-lying clouds are not dangerous; it is their proximity to the ground that makes *them* hazardous for low-flying aircraft. So, inadvertent visual flight into instrument conditions is more a procedural hazard than a weather hazard; however, no matter how you classify it, it is certainly an adverse flying condition, and one that claims many aircraft and lives every year.

How long can a pilot who has no instrument training expect to live after flying into bad weather? The following fictitious hangar story answers this question.

Hangar story

The sky was overcast and the visibility poor. The reported 5-mile visibility looked more like 2. The VFR-only pilot couldn't judge the height of the overcast. His altimeter told him he was at 1,500, but his map revealed there was local terrain as high as 1,200 feet. There could have been a tower nearby because he wasn't sure just how far off course he was. But he'd flown into worse weather than this, so he pressed on. His meeting was very important.

He found himself unconsciously easing back just a bit on the controls to clear those none-too-imaginary towers. With no warning he was in the soup, like flying inside a ping-pong ball (Fig. 5-24). He peered so hard into the milky white mist that his eyes hurt. He fought the feeling in his stomach and tried to swallow, only to find his mouth dry. Now the pilot realized he should have waited for better weather. The meeting was important, but not that important.

His aircraft felt on an even keel, but his compass turned slowly. He pushed a little rudder and added a little pressure on the controls to stop the turn, but this felt unnatural and he returned the controls to their original position. This felt better, but his compass was now turning a little faster, and his airspeed was increasing slightly. He scanned his instrument panel for help, but what he saw looked unfamiliar. It will be all right, he thought, I'll break out in a few minutes, and I'll still make the meeting on time.

He glanced at his altimeter and was shocked to see it unwinding. He was already down to 1,200 feet. Instinctively, he pulled back on the controls, but the altimeter still unwound. The tachometer was into the red, and the airspeed was touching the red line, too.

The pilot was now sweating and shaking. There must be something wrong with the controls, he thought, pulling back only moves the airspeed indicator farther into the

Fig. 5-24. *The pilot found himself flying inside the proverbial ping-pong ball.*

red. He could hear the wind tearing at the aircraft, but view through the windshield gave no clue of his speed or movement.

Suddenly, he saw the ground. The trees were rushing up at him. He tried to find the horizon, but it was at an unusual angle because the airplane was almost inverted. He never made it to the meeting.

This fictitious scenario is based on a study done at the University of Illinois (and published in a Transport Canada flyer entitled, *Take Five ... for safety*). Researchers sought an answer to the question, "How long can a pilot who has no instrument training expect to live after flying into bad weather?" They put 20 student pilots in a simulator. The students flew into instrument weather, and all of them crashed. The outcome differed in only one respect: time required until control was lost. The interval ranged from 20 seconds to 480 seconds with an average time of 178 seconds, 2 seconds short of 3 minutes.

How to get caught

The preceding scenario describes what many pilots would call scud running. Scud running is certainly one of the most common ways to inadvertently enter IMC while flying VFR (discussed in detail in a subsequent subsection of this chapter), but it isn't the only way to find yourself accidentally inside the proverbial ping-pong ball.

Another good way to get caught is to attempt a climb or descent through a hole in the clouds (Fig. 5-25). Such holes are often called *sucker holes* because they sucker pilots in by appearing large enough to accommodate a circling aircraft, but really aren't. Or, they might be big enough in the beginning, but eventually close up above or below you, effectively blocking your way out, before you can get through.

Fig. 5-25. *Flying VFR over a solid bank of clouds is fine until it's time to come down. Descending through a small hole might result in inadvertent IMC.*

A common illusion when climbing through a hole is to think that the cloud tops are lower than they really are; sometimes the clouds are growing as you try to out-climb them. As you climb higher and higher, your airplane's performance decreases, so you climb at a slower rate. If the hole closes up below you, and you can't climb on top, you're suddenly faced with a VFR-into-IMC situation.

When you descend through a hole, you don't have the performance problem, but it is just as easy to overestimate the size hole you need to circle down through. As you reach the fringes of the clouds, the only thing you can do to avoid going IMC is to steepen the turn. Steep descending turns are difficult enough when it is CAVU, but when clouds are nipping the wingtips, the "pucker factor" really goes up. The danger is that you'll enter the clouds with a steep bank and high descent rate. If this isn't a good way to enter an accelerated stall and an eventual inverted flat spin, nothing is.

Two ways to escape

Only two options are available if you inadvertently enter IMC when flying VFR, and both options really require instrument flying skills on the part of the pilot and adequate instrumentation in the cockpit. When these conditions are met, both methods work, although the second is better than the first.

The 180° turn. The first method is the 180° turn. The reasoning is that because you flew into the clouds going one direction, if you turn around and go the opposite direction, you should be able to get out of the clouds. Logically, this should work as long as the maneuver is flown well and conditions don't change. Unfortunately, many pilots have gotten into trouble because they don't fly the turn well, and conditions often change quickly under such circumstances.

The 180° turn is most often successful when it is done *prior* to entering IMC (Fig. 5-26). This is the only time it really works for the noninstrument-rated pilot. Once the view out the windows goes milky gray, the VFR pilot is facing a lifesaving challenge. The IFR pilot isn't in quite as bad a situation, but a lot of things can go wrong.

Fig. 5-26. *The 180° turn is most often successful when it is done prior to entering instrument conditions.*

First of all, the turn must be level, which is not easy. The common error is to climb during the turn because you're worried about the terrain below, if you're already at low altitude. The problem is, when you roll out, you're going to have to descend to get to where you were before. This causes some very anxious moments.

Secondly, the turn must end up at 180°. In the excitement of the moment, it is very easy to undershoot or overshoot the turn. I know because I've done it. Making any turn other than a precise 180° is going to prolong the time in the soup. (Reminds me of the line, "Let's make a 360 and get the hell out of here.")

Another thing to remember is that the turn will take you farther into the low visibility situation, and it is going to seem like a very long time to fly out of it. For example, if you start your turn 30 seconds after going IMC and fly a standard-rate turn that requires 1 minute to turn 180°, it will take another 30 seconds before you fly out of the soup, assuming no-wind conditions. Obviously, wind will complicate the situation, as will an inconsistent airspeed.

The worst case is when the dew point is close to the temperature, and the temperature is dropping. When this happens, you might have the soup forming behind you as well as in front of you. A 180° turn is useless in such a situation. That leaves the second procedure.

The Four Cs. "Climb, communicate, confess, and comply" will avoid the inherent risks of a 180° turn done on instruments and will put you into the controlled environment of the IFR system.

Climb means just that. Level the wings, add full power, and put the aircraft in a climb attitude, eventually trimming for best rate of climb. (You want to trim the aircraft so that you don't inadvertently relax back pressure and allow the aircraft to level out or start descending when you talk to ATC.) Hold heading by using the rudder pedals.

Communicate means call ATC. If you haven't planned beforehand which frequency to use, use emergency 121.5 MHz. Otherwise, call the appropriate ATC facility.

Confess means tell them the truth. Don't pretend you're still VMC and would just like an IFR clearance at their convenience. Tell them you've inadvertently entered instrument conditions, you're climbing through whatever altitude, and you need help. ATC will take it from there.

Your job from then on is to comply with their instructions. You'll probably be directed to fly to and execute an instrument approach at a nearby airport and you're going to need to have the appropriate approach plates on board; therefore, keep them in a handy location.

Which is better? A VFR-only pilot could conceivably cope with the Four Cs if she had some instrument training, but it is going to be a very sticky situation. A VFR-only pilot could conceivably cope with a 180° turn, but it might be even more risky. It is impossible to figure the odds for either case, but neither are good. I really think it is a toss-up as to which procedure is better for the VFR pilot, and this is why I firmly believe that VFR-only pilots should avoid getting themselves into VFR-into-IMC situations in the first place.

For the IFR pilot with all the previously mentioned qualifications and conditions, the Four Cs option is the better bet for reasons already noted. For some pilots, the Four Cs would be hardly more than an inconvenience. ("Oh drat, can't continue VFR any longer. I guess we'll have to climb up and get an IFR clearance. Ho hum.") Other pilots might find themselves working to the limits of their abilities, but when a safe altitude is attained and they are brought under radar control, things should settle down for them, too. Experience makes all the difference.

Hangar story

The destination was Oshkosh, the annual Experimental Aircraft Association bash held every summer in Oshkosh, Wisconsin. Seven people, six of them pilots, made up a three-ship formation of light airplanes that planned to make the trip from Allentown-Bethlehem-Easton (ABE) Airport in Pennsylvania.

The weather was lousy. Visibility wasn't too bad, but there were patches of fog and clouds all around. Fog was even rolling across the end of one of the runways. Poor weather conditions were forecast, and the high-pressure system that was moving eastward was not due to clear things up in Pennsylvania until after the folks wanted to be in Oshkosh. Still, it seemed possible to make the trip VFR.

The three airplanes took off. A 49-year-old woman pilot was flying copilot with her 53-year-old husband in one of the airplanes, their own Piper Cherokee. Both had private certificates and about 350 flight hours each, but neither had instrument ratings.

The formation climbed to 3,000 feet MSL, flying between cloud layers with only occasional glimpses of the farm fields and mountains below. Finally, they came to a large cloud that required about 30 seconds of instrument flying to get through on top. Fortunately, they had had some instrument training. They made it through the cloud without problem, but after this incident they all decided that was enough and began to look for a hole big enough to descend through.

Eventually, the lead airplane found a hole over Selinsgrove Airport, some 30 nautical miles north of Harrisburg and only about 65 nautical miles from ABE. All three airplanes made it to the ground without problem.

The pilots spent three hours at Selinsgrove, gathering weather information and considering options. They finally concluded the weather was too bad to continue VFR over the mountains of western Pennsylvania, and because none of them had instrument ratings, they decided to rent cars and continue to Oshkosh on land. Unfortunately, the closest rental cars were at Wilkes-Barre Airport, about 60 nautical miles up the Susquehanna River. The three took off again.

The weather wasn't much better and the airplanes had to scud-run up the river to Wilkes-Barre, but they made it. Sixteen hours later, traveling by car, they arrived in Oshkosh.

No accident, fortunately, but there certainly could have been. The woman remembers this flight as a "lesson learned."

"The idea is: If in doubt, don't," she says now. "And there was definitely doubt at the beginning of the flight. We should never have started off with the kind of weather we had. We had a very close call, with respect to getting caught in weather. It is one of those things you think about later and say, 'What the hell were we doing?'"

SCUD RUNNING

The term *scud running* probably had its origin in the term "rum running." A rum-runner is a person or ship engaged in bringing prohibited liquor ashore or across a border. Rum running in the United States had its heyday during prohibition. Scud running, on the other hand, is still very popular.

Scud is defined as "loose, vapory clouds driven swiftly by the wind; a slight sudden shower; and mist, rain, snow, or spray driven by the wind." Most pilots think of scud as the thin, misty clouds that hang below thicker clouds and don't necessarily relate it to being driven swiftly by the wind. Patchy fog is considered scud by many pilots. Whatever scud is, everyone agrees that it reduces visibility and makes visual flying more difficult.

A scud-runner is a pilot who flies at low altitude below low clouds in scud. Low clouds and ceilings are often accompanied by low visibility, but sometimes you can have a solid overcast at low altitude and unlimited visibility below. Technically speaking, you wouldn't be flying in scud, but some people might still call you a scud runner.

Scud running is not necessarily illegal, but it can be. Scud running is not necessarily unsafe, but it often is (Fig. 5-27). If the visibility is greater than 10 miles, the ceiling is well defined and above your normal traffic pattern altitude, it is daylight, and you're flying over uncongested areas, most pilots would probably agree the flight can be conducted safely. If any *one* of these conditions changes for the worse—less than 10 miles visibility, ceiling ragged or variable and hangs below normal traffic pattern altitude, darkness, or over a congested area—the safety margin is getting squeezed. And flying might be a violation of the FARs, as well.

Fig. 5-27. *Scud running is not necessarily dangerous or illegal, but it can be both. It should be attempted only by experienced instrument-rated pilots flying low-performance, IFR-equipped aircraft over very familiar routes in daylight with excellent visibility conditions.*

Perhaps the greatest danger with scud running is that conditions might deteriorate as you fly on. Either the ceiling drops lower and lower, or the precipitation that is restricting visibility becomes worse, or both. The pilot starts descending to stay out of the clouds and maintain visual contact with the ground. If the terrain slopes upward, the terrain and clouds will meet at some point. Even if the terrain is level, heavier rain or thicker fog or mist might form a "wall" that drops visibility to virtually zero with little warning. During daylight, this is bad enough; at night, it is 10 times worse.

Scud run: yes or no?

The March 1, 1992, issue of *Aviation Safety* carried an article by John W. Conrad called, "Scud Running." Conrad presented a good case against scud running and, accepting the fact that some pilots were still going to do it, gave some tips on how to make it safer. In the months that followed, a storm of letters from readers either condemned both Conrad and the publication for promoting scud running or praised them for addressing a controversial subject with more than the typical advice: "Don't do it."

Conrad's article and the readers' letters revealed that there are good arguments on both sides of the fence, which is too common in aviation. This makes it difficult to give hard-and-fast rules for scud running. When such is the case, one is left with opinions; therefore, the following compilation of rules for scud running is based upon many opinions of many pilots, including a few notes from Mr. Conrad's article.

Rules for scud running

- A noninstrument-rated pilot should avoid scud running like the plague. Don't even consider it. Your very limited training in instrument flying to get a private license will be of little help. If the weather briefer says "VFR is not recommended," don't go.
- A low-time instrument-rated pilot should avoid scud running like the plague, too. (Arbitrarily, I define "low time" as less than 200 hours of actual or simulated instrument flying.) An instrument rating provides you with the skills to fly in the ATC IFR system, but scud running is a whole different ball of wax. True, you have a better chance of coping with the IFR system than a noninstrument-rated pilot, but your limited experience in the clouds is going to make extricating yourself from inadvertent flight into instrument conditions very difficult.
- An experienced instrument-rated pilot flying an aircraft without IFR instrumentation should not scud run.
- An experienced instrument-rated pilot who is not current in IFR flying (6 hours of IFR flight time, and 6 approaches within the last 6 months) should not scud run.

This leaves us with instrument-rated pilots who have at least 200 hours of actual or simulated instrument flying, who are current, and who are flying IFR-instrumented aircraft. If you're not in this group, don't scud run.

If you are, here are a few more caveats:

- Don't scud run in high-performance aircraft, meaning any aircraft that is not in approach category A, which requires an approach speed of 90 knots, or less. The slower, the better, so that you can fly with a wider safety margin in limited visibility situations.
- Don't scud run at night. Period. Not even in a helicopter.

- Don't scud run just to circumvent the IFR system. If there's a way to legally reach your destination IFR, take it.
- Don't scud run over unknown terrain. Unknown to you, that is. It is easy to get lost even over known terrain, and it is hard to read a map with limited visibility or low visibility. And as good as most sectionals are, they don't portray every obstacle. New obstructions might have been erected since the chart was published, and perhaps the FAA was not properly notified, preventing issuance of a chart change notice in the interim. Best bet: If you haven't flown over the route in good visibility at a low altitude (1,000 feet or lower), within the past two weeks, consider it unknown terrain, and don't scud run over it.
- Don't scud run into an area that is not wide enough to turn around. Imagine flying up a valley or canyon with the ground rising, the clouds descending, and the mountains on each side coming closer and closer. Make your decision to do a 180 before it's too late to do a safe 180.
- Don't scud run into an area that does not give you the option of climbing into the clouds and obtaining an IFR clearance. In other words, if there's nothing above you but cloudy skies, you're all right. But if you're surrounded by mountainous terrain, or the heavy IFR traffic of a terminal area lies ahead, don't go.

Hangar story

The heliport for Helikopter Service in Stavanger, Norway, was located about two and one half miles from the airport. When the airport was below VFR minimums, offshore helicopter pilots commonly flew an ILS approach until visual and below the clouds, proceeding to the heliport on a special VFR clearance. If the weather was too poor, however, they had to land at the airport, and a bus was sent from the heliport to pick up the passengers and crew.

This inconvenienced a lot of people, so pilots had a tendency to push the minimums. The operation was fairly safe for several reasons. The pilots were accomplished at low-visibility, low-altitude flying offshore; they flew the route often; and they knew, or thought they knew, each tree, building, and rock along the way. But this nearly routine scud running was often on the extreme border between safe and unsafe, and the company was lucky no accidents ever occurred.

One fall, new power lines were erected between the airport and the heliport. We saw them doing this, of course, and notices were posted in the flight office. The height of the masts was about 150 feet AGL.

A few months later, I was returning from an offshore flight at night. The weather was crummy, with the clouds close to the ceiling minimum, and visibility was about a mile and a half in rain. Actually, the visibility was our undoing because it wasn't bad enough to make us land at the airport. (It is the borderline cases that will catch you in aviation.)

As copilot on this particular flight, I flew the aircraft on the return leg while the captain did the radio work, kept the flight log, read the checklist, and backed me up on the instrument approach. We saw the approach lights just before decision height and got a view of the runway, but not much else of the airport. Other helicopter flights had returned before us and had flown over to the heliport without any apparent difficulty, so the tower cleared us special VFR to cross the field without our asking for it. Like lemmings, we followed. (When things are borderline, it usually takes a brave soul to say, "Enough is enough," and buck the trend.) We continued across the airport, which was easy because the runway was well lit.

I learned my procedure for scud running between the airport and heliport from one of the older Norwegian captains who had been doing it for years before I started with the company. There were slight variations, but most routines were similar. At the end of the runway, I turned left to 090° which took me directly to a very well-lit service station. The ceiling was about 300 feet, but a bit ragged, so I stayed at 250 feet to keep out of the clouds. Airspeed was set at 70 knots.

I found the service station immediately and thereafter picked up a wide road called "Lowenstrasse." "Strasse" is German for "street," and this particular street was built by the Germans when they occupied Norway during World War II. Originally, it was a high-speed taxiway for fighters between the main airport and another airport that the Germans built in the area where our heliport was now. There were no street lights on Lowenstrasse, but I was able to follow the headlights of a few cars on the road.

The Germans built the airfield where they did because it was surrounded by four hills, and therefore more easily protected than the main airport. Avoiding these hills was my main concern. Fortunately, a local farmer had constructed a large greenhouse right at the corner of Lowenstrasse where it turned left to go through a gap made by the second and third hill. I was looking for the lights of this greenhouse.

With the wind from the west, which it was, the clouds built up in the vicinity of the hills. I had to descend to keep the road in sight and was just on the verge of making a 180° turn back to the airport when I spotted the greenhouse just in front of us. "This isn't good," I said to the captain. He agreed. I glanced at him quickly and noticed he was looking outside just as intently as I was. We both knew that the clouds tended to be a little higher over the small valley where our heliport was located, so we pushed on through the gap.

Usually, once through the gap, we were home free because the lights illuminating the heliport were brighter than any other lights elsewhere in the area. We saw the glow of the lights through the mist and clouds, but we really couldn't see the lights quite yet, even though we were almost close enough to be on downwind. But we felt safe because we were back on our home turf and were so close.

Seconds after we came through the gap, the captain suddenly grabbed the controls and banked sharply to the left 30°, then quickly rolled level again. Surprised, I glanced at him and he answered my inquisitive look with a curt, "We almost hit a tower." I looked past him, out the right cockpit window, and saw the next tower flash by. "That was close," was his next comment, and then he said, "You have control."

As often happens after a close call in flight, you don't have time to stop and think about it. You have to keep flying and attend to the tasks at hand. The heliport came into full view, and I made the landing and taxied to a parking spot. We shut down, unloaded the passengers and baggage, and walked to the flight office. It wasn't until we were inside having a cup of coffee together that the full consequence of what could have happened hit us.

It was odd, actually. We didn't need to say, "We could have been killed out there." We both knew it. One of us said something like, "I think I'm going to raise my night minimums with those towers out there," and the other one agreed. There wasn't much else to say.

I was much relieved a few years later when all offshore helicopter operations were transferred to the airport, effectively ending the Lowenstrasse scud run.

6
Night hazards

"The diurnal death of the world is a slow death. It is only little by little that the divine beacon of daylight recedes from me. Earth and sky begin to merge into each other. The earth rises and seems to spread like a mist. The first stars tremble as if shimmering in green water. Hours must pass before their glimmer hardens into the frozen glitter of diamonds. I shall have a long wait before I witness the soundless frolic of the shooting stars. In the profound darkness of certain nights I have seen the sky streaked with so many trailing sparks that it seemed to me a great gale must be flowing through the outer heavens."

Antoine de Saint-Exupery, *Wind, Sand, & Stars*

THE HAZARDS OF NIGHT FLYING HAVE MORE TO DO WITH THE PERSON INSIDE the cockpit than the environment outside: our inability as humans to see well at night (Fig. 6-1). Outside, things really haven't changed that much, they've just gotten darker. If anything, the flying environment at night is generally better than it is during the day.

At night, the air is usually smoother and clearer, the wind is less, traffic is down, and lighted objects can be seen from greater distances than during the daytime. If the

Discomfort

Glare. Reduced form acuity
and hue discrimination.

Optimum form acuity
and hue discrimination.

Reduced form acuity
and hue discrimination.

Limit for useful form vision.

Absense of color vision.

Outline perception.

Contrast perception.

Light perception only.

Fig. 6-1. *The optimum light level for the human eye occurs naturally on a cloudy day. When overall lighting is brighter, the resulting glare reduces acuity and color discrimination. As it gets darker, our eyes progressively lose their ability to distinguish form and color, with color discrimination going first.*

moon is full and the sky is clear, visibility is nearly as good as it is during the day. Some of my most enjoyable flights have been at night in these conditions. Quite often, after that big heat lamp in the sky sinks slowly in the west, the air settles down, and flying becomes a real delight.

This is not to say that you won't encounter thunderstorms, icing, high winds, frontal passages, and all other manner of adverse weather at night; you might. Some of my least enjoyable flights have been in *these* conditions at night. Sometimes the quiet, stable air of the day is just ripe for the formation of fog and low clouds soon after the sun goes down and the temperature drops a few degrees.

At other times, the air is not stable at all. Sometimes, particularly in the summer, the solar heating that has occurred all day long is so great that the thunderstorms generated reach their peak of maturation after sunset. And large air masses, paying little heed to the "diurnal death of the world," continue their advance upon other air masses, creating cold, warm, occluded, and stationary fronts just as easily at night as during the day.

NIGHT FLYING TIPS

Flying at night requires the same regard for adverse conditions as during the day, but with an additional safety margin factored in to take into account our night vision disability. Here are several things you can do to stack the odds in your favor when flying at night.

Good visibility

Visibility is much more important at night than during the day. You can avoid many of the pitfalls of night flying by simply limiting yourself to times when the visibility is well above the legal minimum. I want at least 10 miles of visibility, with no chance of it going below this in the forecasts (Fig. 6-2). In any visibility that is less than unlimited, it will be impossible to see where Earth ends and the sky begins. One of the biggest hazards at night is that you often can't see clouds ahead of you until you're in them. Scud running at night, as mentioned in chapter 5, should never be considered.

Fly high

Altitude will give you better radio reception, more obstacle clearance, and a longer glide if the engine quits. Fly at least as high as the minimum obstruction clearance altitudes (MOCAs) on IFR charts, even when you are VFR; the minimum en route altitude (MEA) would be better. Don't forget to add 500 feet for VFR cruising altitudes. If you can't make it that high, maintain at least 2,000 feet above the grid altitude on sectional chart. *Grid altitude* is defined as the highest elevation of the ground or obstacles within that particular grid, rounded up to the nearest 100 feet, so it is not the minimum safe altitude, but merely a bare bones minimum altitude.

Fig. 6-2. *Good night flying conditions: full moon, light wind, and good visibility.*

Plot a nonhostile course

Navigating at night might require almost total dependence on electronic navigation equipment because most visual checkpoints, unless they are lighted, won't be visible. In this way, night flying, whether VMC or IMC, is very much like IFR flying. One thing you can do is plot your course to stay as near as possible to airports and nonhostile territory (Fig. 6-3). Stay aware of wind direction, so that in case of an emergency landing you can turn into the wind.

National Severe Storms Laboratory Employees Association

Fig. 6-3. *Plot a course to stay as close as possible to airports and remain clear of hostile territory. This is one way to give yourself an out if an unexpected thunderstorm is encountered at night.*

Light up the airports

Another night flying technique is to light up uncontrolled airports that have pilot-controlled lighting as you fly by them. There's nothing that says you can't do this, even though you might be "wasting" some electricity. A lighted runway makes a great checkpoint and provides an emergency landing strip in case you need it.

On a sectional chart, the symbol "*L" in the airport data block indicates "lighting available on request, part-time lighting, or pilot-controlled lighting." All airports that have pilot-controlled lighting use the same frequency, 122.8, so if any other frequency is listed in the airport data block after the "*L," you can figure that the airport only has lighting on request or part-time lighting. To remove all doubt, look up comments about the airport's lighting in the FAA's *Airport/Facilities Directory* or *Aviation USA* from AOPA.

There are numerous combinations of airport lighting systems, but a simple procedure (as recommended in AIM) will turn on all the systems at the highest intensity. Key your microphone 7 times in 5 seconds.

After the lights come on, you can decrease the intensity of the lights at those airports that have this option by keying the microphone 5 times in 5 seconds, which also lowers the intensity of the runway end identifier lights (REIL), or turns them off at some airports. To get the lights to the lowest intensity at airports with this option, key the microphone 3 times in 5 seconds, again lowering the intensity or turning off REIL systems at some airports.

Pilot-controlled lights stay illuminated for 15 minutes from the most recent time of activation. The lights cannot be extinguished prior to the end of the 15-minute period, except as noted for the REILs. Because all airports that have pilot-controlled lighting use the same frequency, radio-controlled lighting receivers must be set to a low sensitivity when airports are in close proximity; therefore, it might be necessary to fly relatively close to an airport, maybe even right over it, to activate that system.

Turn on the landing light

Flying with the landing light on while in cruise is another good technique. It will do two things for you at night. First, it will make you more visible to other aircraft and any nocturnal birds that might be flitting about.

Second, it will give you an advance warning of clouds and precipitation. If there is any chance at all that the weather can deteriorate rapidly, the flight should be cut short because there are fewer ways to deal with a changing weather situation at night.

In some atmospheric conditions, such as haze or snow, it's probably better to fly with the landing light off. In haze, the light will reflect off the tiny suspended particles in the air and make far-off objects less visible. In snow, it will look like you're accelerating into hyperspace. It's fun to watch for awhile, but very disconcerting after a few minutes. Snow will also make far-off objects harder to see.

Night takeoffs

Takeoffs at night are not much different from daylight takeoffs until you lift the nose off the runway. Up to that point, your visual references are plentiful: lights along the edge of the runway, centerline lights (if installed), the runway illuminated by the airplane's landing light, and other airport lighting. But after tilting your craft skyward, visual references become rather sparse. Taking off from some runways is very much like taking off in instrument conditions. For this reason, the noninstrument-rated pilot needs to be trained and ready to make full use of cockpit instrumentation prior to making any night flights solo.

The climb after the takeoff will be the most difficult because of the nose-up attitude. Once you're leveled off, things generally settle down. On a clear night, the horizon comes into view, as well as the area around the airport. When flying a traffic pattern, it's smart to pay attention to the proper headings and not depend so much on ground reference points. If you fly a standard rectangular traffic pattern, with due corrections for wind, you'll have a much better chance of finding the runway when you turn base.

Approach and landing

Finding a small airport at night, particularly one located in a large metropolitan area, is often a matter of looking for the darkest area as opposed to the brightest area. Often, you will be able see the rotating airport beacon flashing alternately white and green, long before you see the airport proper; in fact, looking for the airport beacon is the best way to find many airports at night. It's amazing how bright shopping centers, service stations, and highways are compared to most airports. Of course, keeping track of your position at all times and using electronic navigation equipment will help you find that elusive airport at night.

Because your visual references are reduced at night, approach and runway lighting become much more significant. It's important not to become fixated on one spot or light, which is quite easy to do when you're concentrating, but maintain a big-picture view of the entire runway so that you can use perspective to provide depth perception and descent guidance. Unfortunately, several visual illusions make this difficult and can cause you to be mistaken about your position and relative speed.

Brighter lights, for example, appear to be closer than dim lights. Just as the brightest stars are not necessarily the closest to Earth, the brightest lights are not always the closest to your aircraft. It could be that they are much, much brighter than the other lights. In fact, you can make a runway seem farther away by dimming the runway lights.

By the way, recall that a pilot can change the intensity of runway lights at airports with pilot-controlled lighting. At other airports, the ATC tower operator or FSS specialist can change the intensity for you. If you think the lights are too bright or too dim, don't be afraid to ask for an adjustment.

If you approach a runway that has an upslope, the runway lights will make it appear that you are higher than you actually are. If the runway has a downslope, you'll appear lower. These are difficult illusions to detect and combat. The best way to overcome any illusions is reliance upon an approach-slope lighting device, if the airport is so equipped: VASI, PAPI, or PLASI. (They are fully described in the AIM.) These lighting systems are designed to provide visual descent guidance information. You should also frequently cross-check the altimeter, vertical speed, and airspeed indicator. Don't be afraid to make a go-around if you find yourself too high or too low.

Another illusion is that the runway lights appear to form a flat plane that is easily mistaken to be the runway surface. Runway lights are usually mounted slightly above the runway surface, so that the flat plane that appears to be the runway surface is actually a foot or two above it. The illusion will cause you to flare a few feet higher than you should. If the runway is wider than normal, your flight path will appear lower than it really is.

Because only nearby objects can be seen clearly, they will appear to move faster than they really are. This will become evident when the landing light beam picks up the runway as you descend closer. Try to avoid fixating on the landing light beam. Bring the whole runway environment into your vision, using it like a big wide-screen attitude indicator. With practice, you'll be able to use your peripheral vision to pick up altitude cues provided by the landing light on the runway surface just before touching down.

Hangar story

In the continental United States, logging night time in the summer is sometimes an inconvenience because you have to wait until after 9 p.m. or so to do it, but in the far north, it's impossible to log night time at all in the summer. When I was flying in that region, after four or five months of daylight-only operations, eventually a lengthy afternoon flight in September would conclude after sunset in night conditions. Even though I had flown in twilight conditions several times during the late summer, maybe even the day before the flight in question, the full nighttime conditions always took me a little by surprise. What clued me in to the fact that it was really getting dark was that I would see lights positioned on an object before seeing the object.

When this happened, I knew I had to consciously compensate for the change in visual cues. I knew that I had a tendency to slow up too soon on an approach at night because I had the perception that I was closing too fast. After slowing up, I'd notice my closure rate was too slow, and I would speed up to compensate. This increase in airspeed usually caused the aircraft to come in too fast, and I'd have to make a quick decrease in airspeed on short final.

Knowing I had a tendency to do this during my first night approaches of the coming dark season, I made it a point to increase my scan of the airspeed indicator during the approach and force myself to avoid decreasing airspeed too soon. After doing this during the first few night approaches, I'd get back in the groove and wouldn't have to think about it as much.

NIGHT EMERGENCY LANDINGS

An emergency landing on unfamiliar terrain at night is going to be extremely risky. Just think how hard it is to choose the least hazardous emergency landing field in full daylight. At night, you probably won't be able to tell the difference between a forest, a lake, a meadow, or a plowed field, not to mention which way the furrows are aligned, unless you have very bright moonlight (Fig. 6-4).

Fig. 6-4. *An emergency landing at night is much more difficult than during the day. Unless you have very bright moonlight, you probably won't be able to tell the difference between a forest, a lake, a meadow, or a plowed field until you get very close to the ground.*

You might be able to see roads and highways, but you won't be able to see power lines, fences, and rock walls. If you know the terrain is relatively flat and obstacle free, perhaps over a desert area, you might be able to survive by letting the airplane settle into the terrain in a nose-up, full-flap controlled glide. Anywhere else, and a landing onto unlighted terrain will probably result in a crash.

One night emergency landing theory says you should head for a dark spot on the ground, the idea being that it will have a good chance of being a field. When you get a few hundred feet above the dark spot, turn the landing lights on to see what it really is. If you don't like what you see, the theory goes, turn the lights off. If you enjoy playing slot machines, you might like to try this method; the odds of winning are about the same.

Perhaps your best bet at night is a lighted highway or large lighted parking lot; both have advantages and disadvantages. A highway will probably have some traffic, the amount depending on the time and other factors. For example, the northbound lanes of the Garden State Parkway in New Jersey usually have bumper-to-bumper traffic from the South Jersey coastal cities to the New York City area every Sunday night during the summer. When landing on a highway, most pilots agree it's better to land with the traffic flow rather than against it, day or night.

The problem at night is that you'll be much less visible to anyone on the road; the good part is that you'll be able to see ground traffic very easily. Landing on the median strip might be a good option in some areas, but this does present other hazards. Some median strips are steeply graded ditches, others are covered in vegetation, others have guard rails. Landing lengthwise in a wide, gently sloping ditch might not be too bad as long as you don't hit an overpass or the U-turn access road that the police are allowed to use. You might not be able to see an overpass or U-turn access road in the dark.

Parking lots might have traffic, too, but it will usually be travelling at a slower speed than the traffic on a highway. If the facility the lot serves is closed, the lot will probably be free of traffic. Parking lot hazards will be light poles and the possibility of wires. The light provided by the poles will be an obvious advantage on landing and most light poles are probably spaced farther apart than the wingspan of most small airplanes, but there will be little room for error. If you can at least get down and dissipate some speed before hitting one, you probably will end up better than if you tried landing in a completely dark area. Wires will be harder to spot. Figure there are wires along the road or street by the parking lot and come in over them. Hopefully, the wires will not be strung from light pole to light pole.

(If you read the three previous paragraphs again, you'll realize I used a lot of qualifying adverbs: probably, maybe, and perhaps. Writers try to avoid using too many qualifiers, but unfortunately—there's another one—one can't write about night emergency landings in any other fashion. There are simply too many improbabilities.)

Not too many years ago, most general aviation pilots considered all night flying an emergency procedure. Today it's considered routine, but many of the risks haven't changed. Let's face it, an engine failure in a single-engine airplane at night is usually going to result in an injury accident, or worse, most of the time. Granted, aircraft engines are much more reliable than they used to be, and most engine stoppages today are due to fuel starvation—an avoidable occurrence, usually caused by pilot error—but anything that is mechanical can break.

I am not saying "Don't fly at night." I am saying "Be aware of the risks, and take all the precautions you can."

RULES FOR THE NIGHT

The United States is comparatively liberal when it comes to allowing flights at night. Many countries of the world do not permit VFR night flying. You must file and fly IFR, no matter what type of aircraft you are flying. Some countries do not allow single-engine aircraft to fly at night under IFR. In the United States, you are allowed to fly VFR at night, you do not have to file a flight plan, and you can fly IFR at night in a single-engine aircraft. The FARs do add a few requirements; pilot flight experience, aircraft equipment, fuel, and weather minimums are addressed.

Night defined

First of all, it's necessary to define exactly what "night" is, according to the regulations. FAR Part 1 *Definitions and Abbreviations* and the *Pilot/Controller Glossary* in AIM define night as:

"The time between the end of evening civil twilight and the beginning of morning twilight, as published in the *American Air Almanac*, converted to local time."

A note appended to this definition in the *Airman's Information Manual*, says:

"Civil twilight ends in the evening when the center of the sun's disk is 6 degrees below the horizon and begins in the morning when the center of the sun's disk is 6 degrees below the horizon."

I admit that I've always found these definitions rather difficult to use in practice. First, I don't have an *American Air Almanac*, and in 20 years of flying I have not looked at one since I was in the Air Force. Second, how does one tell when the center of the sun is 6° below the horizon if you can't see it? I suppose the only way to figure out when the sun is in this position is to time how long it takes for the center of the sun to move from 6° above the horizon to the horizon, and then add the same interval to the time the sun hits the horizon to determine when the center will be 6° below the horizon. For practical purposes, this FAA definition of night seems hopelessly esoteric to me, even though, scientifically, it seems reasonable.

The problem is the FAA expects you to log flight time according to this definition of night, so to do it properly, you should consult the *American Air Almanac* every time you fly at night. The reasoning is that twilight is much closer to daylight than nighttime darkness, which is valid. If your experience is limited to twilight flying, you really aren't prepared for true night flying. So, although the rule makes sense safetywise, it's hard to follow when it comes to writing numbers in your logbook. What should a pilot do?

Sunrise, sunset

Except for the far northern and southern latitudes, twilight typically lasts 15–20 minutes after sunset or before sunrise, so simply add 20 minutes to the sunset time and subtract 20 minutes from the sunrise time to get a good approximation of nighttime. Sunrise and sunset times are published in most newspapers, broadcast on NOAA weather radio, or seen during the local forecast segments on "The Weather Channel," making them easy to find.

Alternatively, note the time when the sun actually goes down, and add 20 minutes to that time; note the time that the sun rises, and subtract 20 minutes from that time. If you're in flight when the sun sets, you'll see the sun longer than someone on the ground. If you use the in-flight sunset time instead of the on-ground sunset time, you might be cheating yourself of a few minutes of night that you could legally log.

Another way to judge when twilight is over and real night starts is the time when you notice the lights on a structure before you see the structure itself. Although this method does not follow the letter of the law, it perhaps complies with the spirit and intention of the law better than the law itself.

Night defined—a second time

While the previous definition of night applies to the logging of flight time, with respect to carrying passengers, the FAA also considers the time from one hour after sunset to one hour before sunrise to be night. Paragraph 61.57 states:

"(d) Night experience. No person may act as pilot in command of an aircraft carrying passengers during the period beginning 1 hour after sunset and ending 1 hour before sunrise (as published in the *American Air Almanac*) unless, within the preceding 90 days, he has made at least three takeoffs and three landings to a full stop during that period in the category and class of aircraft to be used. This paragraph does not apply to operations requiring an airline transport pilot certificate."

This means that you have to log your night landings when it is really dark, not just semidark, which it could be just after the end of or before the beginning of official twilight. Notice also that these landings should be to a full stop, too, not simply touch-and-goes, and the landings must be made in the same category and class of aircraft in which the pilot proposes to carry passengers. It's possible, then, to have hundreds of legal hours of night flying in your logbook, including numerous takeoffs and landings, and still not have the recent flight experience required to fly passengers at night legally.

Hangar story

I was returning with a friend to Merrill Field at Anchorage, Alaska, in late July when I realized that I would be landing after dark. I didn't have the prerequisite three night landings and takeoffs in a single-engine airplane within the last 90 days; I did fulfill the night experience requirement in a helicopter, but that didn't count because a helicopter is a different category and class of aircraft.

I had departed 12 hours earlier. I didn't expect to return so late, but with the sun setting, with one more hour of flight, and with no other suitable airports to land at before reaching Anchorage, the only thing I could do was continue to Merrill. (I might have jokingly asked my passenger to step out of the aircraft because I couldn't legally land with a passenger after dark, but she was a bit nervous about flying anyway and I didn't think she'd appreciate my sense of humor.) My landing was safe, but it could have been better, and I was surprised how much difference the darkness made. I resolved then to make sure I always stayed night current if there were any chance at all that I'd be flying passengers after dark.

Night defined—a third time

Now we have two definitions of night, one for the logging of flight time and another for carrying passengers. As if this weren't enough, the FAA applies another definition with respect to operation of aircraft lights. Paragraph 91.209 states:

"No person may, during the period from sunset to sunrise (or, in Alaska, during the period a prominent unlighted object cannot be seen from a distance of 3 statute miles or the sun is more than 6 degrees below the horizon) (a) Operate an aircraft unless it has lighted position lights. . . ."

Furthermore, you're not even supposed to move an aircraft in dangerous proximity to night flight operations unless the aircraft being moved has the position lights turned on, or is some other way illuminated. "To move" means to taxi, tow, push, or in any other way change position.

When it comes to other aircraft equipment, however, the first definition of night applies: sun's center 6° below the horizon. Paragraph 91.205(c) says you need, in addition to everything required for a day VFR flight: position lights, an approved red or aviation white anticollision light system, an electric landing light (if the aircraft is operated for hire), an adequate source of electrical energy for all installed electrical and radio equipment, and one spare set of fuses, or three spare fuses of each kind that are accessible to the pilot in flight.

Oddly, instrument lights and an extra light source, such as a flashlight, are not required by Part 91, except for large and turbine-powered multiengine airplanes. Needless to say, both of these would be extremely useful in any aircraft at night and are therefore highly recommended. A second flashlight, spare batteries, and spare flashlight bulbs widen the pilot's margin of safety, thwarting a potential operational adverse condition.

When is it really "night?"

To summarize, we have night starting at sunset (with respect to turning on the position lights), night starting at the end of evening civil twilight (with respect to logging of flight time and aircraft equipment requirements), and night starting one hour after sunset (with respect to night flight experience for carrying passengers). It's enough to drive you mad, or make you mad at the FAA.

To be fair, the position-light rule and the passenger-carrying rule are really the only exceptions to the basic definition of nighttime. In every other paragraph I've been able to find that makes reference to night flight, night is used according to the definition, "The time between the end of evening civil twilight and the beginning of morning twilight, as published in the *American Air Almanac*, converted to local time."

Requiring pilots to refer to an obscure almanac before or after every night flight might be unrealistic, but at least the FAA is more or less consistent. Where do you find an *American Air Almanac*? When I asked the owner of the airport where I tie down my airplane if he had the almanac, he didn't know what I was talking about. When I told him I was looking for sunrise/sunset times, he said, "Oh, no problem," and showed me an old, dog-eared sunrise/sunset chart of unknown origin that he kept by the cash register.

I suspect most FBO operators and pilots make do in some similar fashion. The Aircraft Owners and Pilots Association's reference guide, *Aviation USA*, contains a sunrise and sunset table with the times for six major airports across the country with two dates each month. The text explains how to extrapolate to get a close approximation of the sunrise and sunset times for any location on any day of the year.

Be careful when referring to newspapers. I carefully followed the times published in my local daily newspaper for a month and found several inconsistencies, such as sunrise being 5 minutes later one day than it was the day before, when it should have been 2–3 minutes earlier. It was obviously a mistake, but if you only looked at the time given for that day, there would have been no way of knowing it was wrong.

More rules for the night

The basic definition of night applies to nighttime fuel and weather minimums, too. The night fuel-reserve requirement is simple enough. Instead of a 30-minute fuel reserve required during the day, you must have a 45-minute reserve at night. Most pilots probably carry a longer reserve than this both day and night, but it is all that the FAA requires.

The visibility requirement at night is 3 statute miles in all airspace below 10,000 feet MSL. Above 10,000 feet MSL (and more than 1,200 feet AGL), you need 5 statute miles visibility. The requirements for distance from clouds when flying below 10,000 feet are 1,000 feet above and 2,000 feet horizontally; above 10,000 feet, you need 1 mile horizontally.

These are, of course, only the minimums. The wise pilot will have self-imposed minimums that exceed 1,000 and 3 at night, unless instrument rated and current, flying an IFR-equipped aircraft, and mentally prepared to file IFR at any time.

THE EYES AT NIGHT

Flying at night would be more of a delight if the eyesight of the human species extended into the infrared range. Because it doesn't, the only way to get improved night vision is to supplement your own eyes with infrared night-vision goggles, a *forward-looking infrared* (FLIR) system on the instrument panel, or some other vision-enhancing device (Figs. 6-5 and 6-6). Unfortunately, these are still outside the realm of most

Fig. 6-5. *Someday all aircraft designed for night flight will have some sort of vision-enhancing equipment, such as this forward-looking infrared (FLIR) display in a police Bell 206 helicopter. The white, ghost-like image on the FLIR screen is not a reflection, but the thermal image of a person standing in front of the helicopter.*

Fig. 6-6. *The FLIR ball is mounted on the helicopter's belly, behind a searchlight. A FLIR thermal imaging system is essentially a video camera equipped with a special lens and detector designed to "see" radiation in the infrared portion of the electromagnetic spectrum instead of processing an image in the visual portion.*

civilian pilots and the FAA has yet to come to a definitive decision for their use during nonmilitary applications.

Eventually, night-vision goggles, FLIR, or some other vision-enhancing device might become part of everyone's standard flight kit for night flights. I'll be more bold with my prediction and say that someday they *will* become standard for all night flights because the devices are that good. For many years to come, most of us will have to cope with the basic human eyeball, standard model, two each.

As good as our eyes are during the day, their ability to see objects in darkness is extremely limited. We must even be careful when using the term "darkness," because in the total absence of light, our eyes see nothing at all. What we are able to see at night is due to whatever available light is present, be it from the stars, the moon, or man-made devices. The dimmer the light, the less we see.

Rods and cones

The physical construction of our eyes is the reason for our limited night vision. After light passes through the lens, images are focused on the retina on the inside back wall of the eye. The retina is a mosaic of hundreds of thousands of nerve endings that function as receptors. These nerve endings are either cones or rods, both of which are named for their shape. Each retina has approximately 6–7 million cones and more than 100 million rods. The cones are connected individually and the rods are connected more collectively to cells that terminate in fibers. The fibers collect into the huge bundle of fibers known as the optic nerve.

The cones are used primarily for daylight vision, activating when the light is equivalent to moonlight-strength or greater. Only cones can distinguish color. Everything we look at directly comes to focus on the fovea, an area near the center of the retina that is smaller than a pinhead. All our finely detailed seeing is done in the fovea. The fovea is composed only of cones, which explains why we can see small things best in brightly lit conditions.

Beyond the fovea and a slightly larger area that surrounds it called the macula lutea, the cones dwindle and the rods take over. The rods are much more sensitive to light than the cones, but cannot distinguish colors or fine details. We see best in dim light not by looking directly at an object but to one side of it, so as to focus the light on the rods on the edges of the retina, instead of the cones in the fovea. In essence, we must sacrifice detail in order to see anything at all in dim light. You can demonstrate this night blind spot by staring at a faint star on a clear night. After a few seconds, it will disappear from your vision. Look away slightly, and it will come back into view.

The cone-filled fovea accounts for one blind spot, which is there only in low-light situations. The other blind spot is at the place where the optic nerve enters the back of the eye. This blind spot is with us all the time, day and night. Fortunately, the brain fills in the background detail around the spot for us so that we aren't aware of it. You can find the optic nerve blind spot by looking at a small black dot on a white sheet of paper with only one eye, then moving your eye slowly toward your nose while keeping it focused on the paper. The dot will enter the blind spot caused by the optic nerve; after a second or two, the brain will stop filling in the background, and the dot will seem to disappear.

Visual purple

Our night vision is further enhanced by a substance that is produced by the rods. The substance is called *visual purple*. It comes into play when your eyes slowly adjust to low-light situations (Fig. 6-7). This dark adaptation process can take from 30–45 minutes for the eyes to become adapted, but it takes only a second or two of bright light to wipe out the positive effects of visual purple. Recall the concerns about lightning that are outlined in chapter 5.

Use of red lights in the cockpit will help maintain night vision, but this is not as important as once thought. Dimming the instrument lights to the lowest readable level is just as good. (Although Vitamin A is needed by the rods to produce visual purple, there is no conclusive evidence that extra Vitamin A in your diet will improve night vision. A lack of it, on the other hand, can cause a decrease in night vision capability.)

Oxygen deficiency causes night vision to deteriorate. At 4,000 feet, your night vision is about 5 percent less effective than at sea level; at 6,000 feet, 20 percent less effective. Heavy smokers can lose 40 percent of their night vision at 6,000 feet due to the reduction of their blood's ability to carry oxygen.

INSIGHTS TO REMEMBER

To summarize, there are three important things you should remember about your night vision:

- You are not going to be able to see unlit objects at night as well as you can in the daytime, no matter how good your vision is.
- You won't be able to see an object better at night if you stare directly at it, as you can during the day. In fact, you'll make it harder to see. Keep your eyes moving to avoid the blind spot caused by the cone-packed fovea.
- Give your eyes 30–45 minutes to fully adapt to the darkness, and avoid bright lights to keep them adapted.

Fig. 6-7. *Visual purple enhances our night vision in low-light situations, such as on moonlit nights.*

7
Terrestrial hazards

"O Beautiful for spacious skies,
For amber waves of grain,
For purple mountains majesties
Above the fruited plain!"

Katharine Lee Bates,
America the Beautiful

IF WE ALL LIVED IN GREAT CITIES FLOATING IN THE SKY, THERE PROBABLY would be a lot fewer aircraft accidents. Aircraft and pilots do quite well while en route, except for the occasional tangle with weather or other aircraft and the even less frequent structural breakup.

Unfortunately, skyborne cities are found only in science fiction stories, so we must contend not only with the hazards of the air, but also with the hazards associated with the surface of the earth. Obviously, some surfaces are more hazardous than others: mountains and large bodies of water, for example. And man-made obstructions, such as power lines, make even the most featureless plains a hazard.

Finally, another hazard to aircraft is included in this chapter: birds. This hazard posed a bit of an organizational problem for the book. Birds obviously aren't a weather hazard, and they aren't really a surface hazard. You might think it stretches the meaning of "terrestrial" to include birds, but my dictionary defines terrestrial as, "of, or relating to, the earth or *its inhabitants*." Birds are inhabitants of the earth; therefore, they fit in this chapter better than anywhere else.

MOUNTAIN FLYING

"Flee as a bird to your mountain."

Psalms II:I

The central problem of mountain flying can be expressed in two or three words: "density altitude" or "high density altitude." Everything else that has to do with mountains, including winds, long takeoff rolls, turbulence, whiteouts, decreased engine performance, downdrafts, mountain waves, and the like, is either directly caused by, or exacerbated by, high density altitude. The reason is very basic.

Nonturbocharged reciprocating engines lose power as density altitude increases. A Cessna 172, for example, attains only about 50 percent power at 8,000 feet density altitude. With less power, takeoff runs are longer, and climb performance deteriorates. With less engine power, downdrafts are harder to overcome, turbulence takes longer to fly out of, and cloud tops might be too high to fly over.

For example, a 150-horsepower Cherokee 140 crashed near Monarch Pass, Colorado, (elevation 11,200 feet) after encountering a downdraft that was so severe the airplane was unable to stop descending even at full power. It's not surprising that aviation death rates in the mountain states are at least twice that for the United States as a whole.

This has not gone unnoticed. The General Accounting Office released a report in early 1994 entitled "The FAA Can Better Prepare General Aviation Pilots for Mountain Flying Risks." The office identified a select group of mountain airports that had an accident rate 155 percent higher than a select group of nonmountainous airports with similar numbers of flight operations. GAO recommended that the FAA issue guidance to pilot by identifying mountainous airports that present unique challenges, and develop recommended routes for takeoff and landing at these airport. The report also stated that general aviation accident rates are nearly 40 percent higher in the western states that are designated as mountainous than for the other continental states.

Mountain flying is something that really must be experienced firsthand. You can read about it, but it's impossible to understand what it's like until you actually fly in a mountainous region. If you haven't learned to fly in mountainous areas from pilots who have done it, don't try it without getting some firsthand advice. Books can't provide mountain flying specifics, which are abundant (Fig. 7-1).

Pilots who fly in one area on a regular basis know that weather can be very localized. A small river or lake can produce a thick fog bank while the surrounding area is clear. A single smokestack can produce a wide swath of smog that alters the normal traffic pattern at an airport. A few low hills can generate so much orographic turbu-

Fig. 7-1. *To really learn about mountain flying, you must experience it firsthand. Books can give you good general information, but before you head for the hills, seek the advice of local mountain pilots who can clue you in on the specifics of the area, like typical weather patterns, radio irregularities, and the best places to get fuel and coffee.*

lence in some wind directions that an approach to or takeoff from a particular runway becomes extremely hazardous.

In the mountains, the frequency of similar or identical disturbances is doubled, tripled, and even quadrupled. Mountain weather is actually composed of thousands of small-scale weather systems that can result in wide variations over relatively small distances. Mountain formations create their own weather.

What this means for the pilot flying a low-powered, or even a medium-powered, airplane is that mountains will offer weather that probably won't be forecast. This also means that the pilot will have fewer options available to cope with this weather. In other words, the margin for error is narrowest precisely at the time when you'll need the widest margin that you can get.

The following information will help make your margin as wide as possible. Supplement it with advice from pilots who are familiar with flying in the mountains that you plan to navigate through.

AIRCRAFT PERFORMANCE

Knowing the performance capability of your aircraft harps back to planning again, but it's more critical in the mountains than anywhere else. Check out the performance charts for your airplane before you go; you'll be surprised. A common misconception is that a larger airplane has a better rate of climb than a small one. Most do, but only if you carry less than maximum gross. Fill all four seats in many four-seat airplanes, add fuel and baggage, and you might actually have less rate of climb at higher altitudes than a two-seat trainer loaded to maximum gross weight.

Know the airplane's service ceiling. Service ceiling is defined as the altitude at which an aircraft, at maximum gross weight, will still be able to climb at 100 fpm at full power. A 100-fpm climb is not very much; at that rate it will take you 10 minutes to climb 1,000 feet, assuming, of course, that your airplane can maintain that rate.

The problem is as you climb higher, the engine horsepower continues to decrease and the rate of climb decreases, too. So, if you're already at your airplane's service ceiling, don't plan on being able to climb at 100 fpm much above that. On the other hand, the airplane's gross weight will decrease as you burn off fuel; theoretically you should be able to maintain some rate of climb above the service ceiling, but most flight manuals don't provide you with enough charts to make it simple to figure what that rate will be. This is a case of aeronautical WYSIWYG (what you see is what you get).

The point: Don't plan to push your airplane much higher than its service ceiling. If your route takes you over mountains that rise above your airplane's service ceiling, figure you'll have to go around them. A rate of climb that is 100 fpm or less is a rather narrow margin when encountering downdrafts in the mountains.

Fudge factors

If the book says your airplane can take off at maximum gross weight with 6,000 feet of runway under certain conditions, add a 25-percent fudge factor to that when flying in the mountains (Fig. 7-2). Fifty percent will be even better. Remember, the figures in the pilot's operating manual were achieved by an experienced company test pilot flying a new airplane with an engine at the very top of its performance. Manufacturers want their aircraft to look as good as possible when compared to other aircraft; they do everything legally possible to make those figures as good as possible. Your airplane is going to have less performance than the test airplane. In most cases, you will not be as experienced as the company test pilot. Finally, remember that the test pilot probably made several attempts before achieving the best numbers to put in the flight manual. You only have one chance to make that marginal takeoff.

The AOPA Air Safety Foundation recommends that you don't even consider taking off unless the flight manual says you can maintain at least a 300-fpm rate of climb on departure. This will give you a built-in fudge factor to help you overcome unexpected and unusual winds and the probable less-than-optimal performance of your airplane. Another rule of thumb states that the takeoff and landing distance for a nonturbocharged piston airplane increases by 25 percent for each 1,000-foot increase in density altitude.

Probably the most important fudge factor is fuel. Don't skimp here. If you can't plan to take at least a 1-hour reserve, then seriously consider landing at an intermedi-

Fig. 7-2. *Give yourself a hefty margin of error when determining takeoff and landing distances from mountain airports. It doesn't hurt to add 25–50 percent to the figures in the performance charts.*

ate airport to refuel. A 100-percent reserve is not out of the question. There are too many variables in the mountains than can cause you to use more fuel than you planned. The wind speed can be higher than expected. Wind direction can change in a short period of time from tailwind to crosswind to headwind. You might have to decide between a strong headwind up high or excessive turbulence down low. Your primary route might be shrouded in unexpected clouds, causing you to take a longer route. A mountain pass might be completely obscured and impassable, necessitating a return to your departure point.

If you check the charts and find the runway you want to use is shorter than you'd like, reduce weight. Know how much you load into the aircraft by using a bathroom scale to weigh everything you put in it except fuel; don't forget to add the weight of the fuel. Don't leave your survival equipment behind to reduce weight. If you have a four-place airplane, don't figure on carrying four people with baggage unless the airplane has a turbocharged engine; a turbocharged engine still might not be enough. You might find it necessary to make two or more trips between airports in the mountains to carry the same number of passengers and baggage that you can carry on one flight between two airports at lower altitudes. Don't fight it. If that's the safest way to do it, then make the extra trips and enjoy the flight time.

Finally, give yourself plenty of time. Mountain flying is the epitome of the adage: "If you have time to spare, go by air." Don't force yourself to make a flight or continue a flight in marginal conditions because you have to "be there on time, or else." Give yourself a day or so to be able to wait for better conditions before you take off. On the day of the flight, give yourself a few hours to spare in case you have to make a diversion. You don't want thoughts of "getting there" to cloud your decision-making in the mountains.

Mixture control

Most low-landers are accustomed to using full rich mixture for takeoff; however, takeoff at higher altitudes requires a lean mixture to get the most horsepower possible from the engine.

The recommended technique is perform the standard runup, then advance the throttle to full power to activate the effect of the enrichment valve. While at full power, lean the mixture. For aircraft without an EGT gauge, simply lean the mixture for highest static RPM, then enrich slightly. If you have an EGT gauge, lean to peak temperature, then enrich the mixture to about 100°–150° on the rich side of peak.

It's important to lean the mixture when operating at excessive density altitudes, not only to get maximum available power from the engine, but also because it only takes a couple of minutes at full rich for the spark plugs to foul. This can cause further loss of power and even engine failure.

Supplemental oxygen

The rate at which the lungs can absorb oxygen depends on the partial pressure of oxygen in the air. At sea level, the total standard pressure is about 14.7 psi. Because oxygen constitutes about ⅕ (21 percent) of the atmosphere, the partial pressure of oxygen at sea level is around 3.0 psi. This is what your lungs are used to, unless you happen to live at higher altitudes.

Because air pressure decreases as altitude increases, oxygen pressure also decreases. Lower oxygen pressure means the lungs absorb less oxygen. The symptoms and effects of oxygen deficiency, called *hypoxia*, are a feeling of exhaustion and fatigue, blurred and tunnel vision, mental confusion, hyperventilation, apprehension, nausea, vomiting, dizziness, hot and cold flashes, euphoria, belligerence, numbness, tingling, poor judgment, and muscle incoordination. Flying at altitudes above 10,000 feet for a prolonged period of time will subject a person to hypoxia (Fig. 7-3).

Be especially aware of hypoxia hazards at night, when the symptoms can occur at altitudes lower than 10,000 feet. Good vision is a crucial element of night flying. The slightest visual deterioration might go unnoticed in daylight, but prove dangerous at night.

Most people don't exhibit all the symptoms of hypoxia, and some people tolerate slight oxygen deficiencies better than others. All other things being equal, if you live at high altitude, you probably can tolerate oxygen deficiency better than someone who lives at sea level. The use of some drugs, including common over-the-counter medica-

Cessna Aircraft Company

Fig. 7-3. *Mountain flying often requires cruise altitudes above 10,000 feet. At these altitudes, hypoxia can become a serious problem for most people. Use of supplemental oxygen above 10,000 feet is a good idea.*

tions, any alcohol in the bloodstream, fatigue, stress, and cardiovascular disorders will exacerbate the effects of hypoxia. Smokers and overweight people also tend to have more severe hypoxia problems.

Regardless of where you live or your physical condition, the effects of hypoxia are cumulative; the longer you deprive your body of oxygen, the more symptoms you'll get. You'll become unconscious if you deprive your body of oxygen long enough.

FAR 91.211 sets forth the requirements for oxygen usage when flying. Above 12,500 feet MSL, up to and including 14,000 feet MSL, the flight crew of an unpressurized aircraft is required to use oxygen for that portion of the flight that is more than 30 minutes long. Above 14,000 feet MSL, the flight crew is required to use oxygen for the entire time. Above 15,000 feet MSL, everyone on the airplane must be provided with supplemental oxygen available.

The Air Force regulations are even stricter. When minimum en route altitudes or an ATC clearance requires flight above 10,000 feet MSL in an unpressurized aircraft, the pilot at the controls must use oxygen. When oxygen is not available to other occupants, flights between 10,000 and 13,000 feet must not be longer than 3 hours, and flights above 13,000 feet are not authorized.

Other sources recommend using supplemental oxygen whenever you fly above 10,000 feet. Considering the debilitating effects of hypoxia and the relatively low cost of carrying oxygen, this seems like a very good idea.

Weather you can always count on

Mountain weather is composed of thousands of small-scale weather systems of varying air density, humidity, and temperature lapse rate. Actually, there are three separate climates in the mountains: valley, slope, and the free air above. The mixing of these climates ensures weather variety throughout the year.

Don't rely solely on broad-based weather forecasts and reports. You need nitty-gritty details about the local weather to fly in the mountains. Talk to local pilots and meteorologists. (One advantage of the often long distances between airports in mountainous areas is that flight service stations are still available at many smaller airports.) Find out which areas are generally good and which ones should be avoided.

Sometimes it's impossible to pick the best route simply by looking at a chart. Sometimes the mountain pass that seems to be the most favorable isn't. Call your destination airport to get an on-site evaluation of the weather. Call airports along your route to fill in more details.

Be aware that clouds have a tendency to cling to mountain tops and ridges (Fig. 7-4). Mountain peaks are often adorned with banner clouds, which are formed by the sudden cooling of air as it is thrust over the summit. Such clouds are easily circumvented and announce the presence of higher level winds and turbulence.

Mountain passes offer an enticing route when the surrounding ridges are in the clouds, but be sure to check your chart for the maximum terrain or obstruction eleva-

Fig. 7-4. *Clouds have a tendency to cling to mountaintops and ridges.*

tion within the pass. Often the opening of the pass will be in the clear, but the middle of the pass will rise up to meet the overcast. Merrill Pass in Alaska was well-known for catching pilots. To fly through it, you have to turn about 45° right, then almost 90° left, and finally 45° right again. After the 90° turn, there isn't enough room to turn around if the other side of the pass is obscured. The broken fuselages of numerous airplanes on the floor of the pass attests to this fact. If the chart says the floor of a pass rises to 3,200 feet, and the cloud base at the mouth of the pass is 3,000 feet, you should find another route through the mountains.

Glaciers add their own particular twist to mountain weather (Fig. 7-5). A glacier is a large body of ice and snow that covers a land surface year round. The larger the glacier, the more it will affect local weather. The weather and climate of Greenland, for example, is totally dominated by the glacier that covers nearly the entire island to an average depth of 1,000 feet. Because the surface temperature of a glacier is below freezing, warm, moist air that flows over the glacier will immediately condense into fog or low clouds. Depending on the surrounding terrain and prevailing wind, these clouds can be pushed into surrounding valleys when every other area is clear.

Glaciers also create their own winds by cooling the air immediately above the surface. When the ambient air is above freezing, this layer of cool air right above the glacier will flow downhill. The higher the ambient air temperature, the faster the air di-

Fig. 7-5. *The cooling effect of the air directly above glaciers can cause strong local winds and create low clouds and fog that might move into surrounding valleys.*

rectly above the glacier will move, often creating an area of considerable wind shear and turbulence.

Although they rarely combine into squall lines, mountain thunderstorms are to be avoided for the same reasons as any thunderstorms that form over land or water. High air-mass humidity and convective factors help these orographic thunderstorms rapidly grow to impressive size and heights. A thunderstorm stubbornly spanning an entire valley is a common sight in many areas. If you have enough fuel on board, you can probably circumvent such a storm. Otherwise, land and wait for it to dissipate or move on to another valley.

Hangar story

"I was flying along when suddenly I saw a cloud, a big cloud. Out West, they speak of things differently than back East. When they said 'scattered conditions,' they apparently meant one big cloud."

The speaker is a feisty grandmother who has a commercial certificate and an instrument rating. At the time of the incident, she was 61 years old and had about 700 flight hours.

Flying a Piper Arrow solo from Easton, Pennsylvania, to Great Falls, Montana, one July, she had filed and flown IFR, mostly in VMC conditions, until she arrived in Sheridan, Wyoming. She likes to fly on an IFR flight plan because it gives her the feeling that she isn't alone, that someone is watching her flight from below. She has another good habit; she always traces her flight path on a VFR sectional as she flies along, whether IFR or VFR. "I look on the chart to see what's under me, even if I can't see it," she explains.

She landed at Sheridan, which was "severe clear," ate lunch, and checked the forecast for the last leg to Great Falls. Unlimited visibility and only scattered clouds were called for, so she decided to continue on a VFR flight plan while following the victor airways. She chose a safe altitude by checking the minimum safe altitude on the VFR sectional, which was 8,500 feet MSL, and applied the hemispherical rule to the westerly heading; she selected 10,500 feet MSL.

"It took me an hour to climb up there, but that was all right," she recalls. "Then I saw that cloud." She isn't sure what kind of cloud it was, except that it wasn't a thunderstorm, but remembers it as being huge. "Too high to fly over and so wide it would have taken me 200 miles off my course. I didn't want to fly under it because of the mountains." After a few minutes mulling over the problem, she called Salt Lake City Approach and asked for an IFR clearance, which was granted immediately. ATC told her to climb to 11,000 feet.

"No problem, I thought. I flew into the clouds at 11,000 and all of a sudden I'm getting cold. Now, when I left the ground it was about 86°" she recalls. "I put on a sweater, and then it dawns on me—I'm at 11,000 feet and I'm in a cloud. I looked at the thermometer and thought uh-oh, and right then I started picking up rime ice."

Because she had been following her track on the sectional, she knew exactly where she was over the terrain. "I called Salt Lake, told them I was picking up ice, and requested a descent. I could not see out the windshield. They said, 'You can't go down, you have to go up.' I told them I knew I was near an air force base and the base could pick me up on radar and that I had to go down. It was one of the first times I really asserted myself." The controller sounded upset and went off the air for a moment, but came back with a clearance to descend to 8,000 feet and contact the air force controller.

"At about 9,000 feet, I started going through rain and about 8,500 the ice slid off the windshield. I continued on to Great Falls with a lot of help from some very nice people."

The pilot learned a valuable lesson. "It never dawned on me, being from the East and since it was July and hotter than hell on the ground, that going into that cloud I would pick up ice. When I got to Great Falls, the people there told me very few of them use IFR clearances because usually you can see 120 miles and if you're IMC at those altitudes, you're usually in ice."

Mountain winds

The best way I know to visualize wind in the mountains is to think of it as water. Try to imagine a giant river of water flowing over and around mountain ridges and peaks, into valleys, and over cliffs (Fig. 7-6).

On the upwind side of a ridge, the air will be smooth and the rising slope of the terrain will cause the air to rise. If you fly high enough over the ridge, the wave of air will push you up smoothly and steadily, without any increase in power. As the wave flows over the ridge and down the other side, you will ride downward with it. This is fun in a glider, but if you're trying to maintain a constant altitude, you'll have to decrease power before reaching the ridge and add power after passing it.

It's impossible to say how high is "high enough" because it depends on wind velocity, wind direction, ridge height and shape, and aircraft performance. Usually, the stronger the wind, the higher you'll need to fly to stay in the smooth updrafts and downdrafts of a mountain wave.

Lenticular clouds are a good visual illustration of this river analogy. The clouds start to condense as moist air on the windward (upwind) side of a mountain ascends up the slope. The maximum thickness of the clouds is over the summit. As the air flows down the leeward side of the slope, it dries, and the cloud dissipates, forming the distinctive lens shape of the lenticular cloud. Lenticular clouds are like waterfalls in that they are fixed in one place, but are composed of constantly changing material.

If you fly "too low" over a ridge, you'll get caught in the turbulent air on the other side. Imagine the Niagara River as it reaches the Niagara Falls. The last thing you want to do is fly into the air cascading down from the top of the ridge.

The precise area of this turbulent air on the downwind side of a ridge is difficult to determine by just looking at a ridge. With light and medium winds, the turbulence will

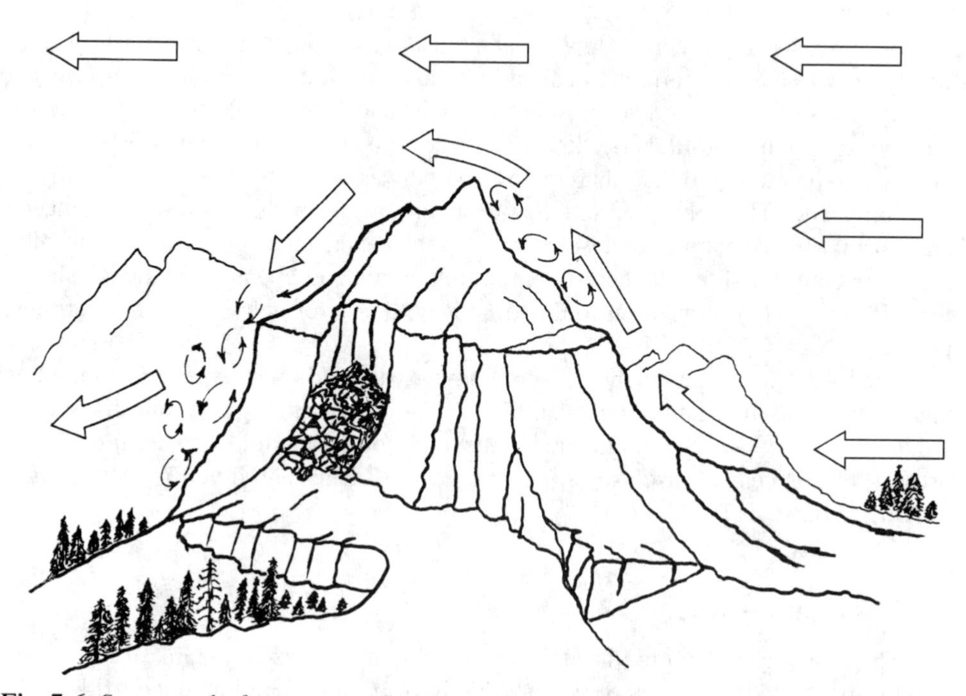

Fig. 7-6. *Strong updrafts are created when wind at the surface encounters a mountain and moves upward. Strong downdrafts are created when the wind reaches the mountaintop and flows downward. Any irregularities in the shape of the mountain will create areas of turbulence.*

usually be right up against the side of the slope. In stronger winds, however, there might be an area of relatively calm air that is protected by the ridge. This area is usually too close to the ground for normal flight, but sometimes it can be useful.

(A rescue helicopter needed to land at a small farm on the downwind side of a high ridge in Iceland. The wind was blowing about 30–35 knots, and there was considerable turbulence in the approach path into the wind. By accepting a slight crosswind and coming in at a shallow angle below the ridge line, the pilot was able to avoid the turbulence coming off the top of the ridge and make a safe approach and departure.)

Heat from mountain slopes that face the sun complicates the turbulent effects of the wind coming over the ridge. Not only can this cause the formation of clouds, it will also cause considerable turbulence from the ground up. Clouds will be a good clue to expect turbulence on this side of the ridge. These clouds usually don't rise up too high, especially in early morning, so they can be flown over without difficulty.

Another hazard of mountain winds is the venturi effect. The venturi effect/Bernoulli principle also lifts an airplane's wings. As a quantity of air moves through a restricted place, such as a mountain pass, the velocity of the air increases, and its pressure decreases. Imagine a giant funnel concentrating the wind though a small opening between two ridges. The wind speed in a pass or valley might increase to such a degree that a slow aircraft might not be able to make any acceptable headway in the opposite direction. In one of his books, Richard Bach tells about trying to fly an antique biplane against the wind through a mountain pass in California. He finally gave up and did a 180 when he saw the wind was so strong that he was actually moving backwards over the ground.

When you fly along a valley, always fly on one side of the valley instead of straight down the middle (Fig. 7-7). This will give you a wider area to turn around in if you need to. It will also make it easier to spot power lines crossing over the valley. Tip: Look for the support structures because most power lines are not marked with visibility devices. Although you might think it's better to avoid power lines by flying over the middle of a valley because the lines droop down, they are very hard to see and you never know how high they are. If you stay above an imaginary straight line between the tops of two consecutive support structures, you can't fly into the lines, which normally droop between structures. Beware: If the structures go upslope, the lines will rise from the top of the lower structure; steer away from upslope structures to reduce the risk of a wire strike.

Fig. 7-7. *When flying in a mountain valley, stay along one side, so you'll have maximum room available to turn around. Because of the friction of the valley walls, wind speed will usually be slower on the sides than in the middle of the valley. Excessive turbulence can also be found near the bottom.*

Choosing which side of a valley to fly along depends on the wind direction. Always fly on the downwind side so that you're taking advantage of the updrafts caused by the wind moving up the slope. This will also give you a shorter radius turn into the wind if you need to make a hasty retreat. Flying on the upwind side of a valley will subject you to continuous downdrafts and turbulence.

When crossing a ridge from the downwind side, fly at a 45° angle to the ridge with as much clearance over the ridge as you can get. The lower you are, the more turbulence and downdrafts you'll be fighting. Although a 45° angle will place you in this turbulent region longer than flying straight at the ridge, it gives you an out if the downdrafts get the better of you. When this happens, you only need to make a 90° turn away from the ridge to get out of there.

On the subject of downdrafts, if you get caught in a prolonged downdraft, avoid the temptation to pull up the nose to regain altitude. Instead, push the nose over to gain airspeed and climb performance to minimize the amount of time in the downdraft. At some point above the ground, the downdraft will have to dissipate horizontally, and the wind velocity will decrease, so it's unlikely that a prolonged downdraft will carry an airplane all the way to the surface.

You have a better chance of flying out of the downdraft if you're flying at cruising speed when you level out at treetop level. Controllability is significantly diminished when the airplane is mushing down with the nose up and the airplane about to stall.

It's also preferable to be heading away from the slope toward lower ground than up the slope if you encounter a downdraft; therefore, don't wait too long to turn away from a ridge if you think you're going to have a problem flying over it.

Best times to fly

One or two days of flying in the mountains will reveal that the best times to fly are in the early morning hours or late in the afternoon. The radiation from the sun on clear days on the "sunny side" of the mountains produces turbulence and thunderstorm-producing updrafts (Fig. 7-8). When I lived in Colorado Springs, it was almost possible to set your watch according to the thunderstorms that appeared nearly every day in the summer at about 3 p.m. In Alaska, Utah, and Norway, I observed the same phenomenon, although the timing varied according to local conditions. If the rain showers didn't get you, the turbulence did.

The cooler temperatures of the morning and afternoon also benefit aircraft performance because density altitude is mainly a function of temperature.

If you have a choice, fly the dawn patrol instead of the dusk patrol. Calm and clear conditions will normally last from daybreak to at least noon. Turbulence and cumulonimbus activity will increase steadily until early evening. You might get two or three hours of smooth flying in the evening until dark, but if the surface heating has been particularly vigorous, thunderstorm activity might continue well past dark. In the mountains, it's generally better to fly with the hawks in the morning than the owls at night.

Be afraid of the dark

I have a fair amount of night flying experience, much of it in and out of the fjords of western Norway, in the mountains of Alaska, and throughout Iceland. I was flying

Fig. 7-8. *Afternoon "thunder-bumpers" form nearly every afternoon over the Alaska Range during the summer months*

modern, twin-engine helicopters that were fully equipped with IFR gear, area navigation equipment, and radar. Flying over mountains at night is not something I'd want to do in a single-engine airplane.

Even on clear nights, mountains are hard to see. Surface lights are few and far between. It's impossible to tell when clouds are clinging to the peaks and fog is forming in the valleys. IFR equipment is essential, but even with it, I question the safety of flying a single-engine aircraft in the mountains at night. What do you do if the engine quits, or if it has even a partial failure, or if you have an electrical failure? There just isn't any safe way out.

Therefore, even though it is legal to fly VFR at night in a single-engine aircraft over mountainous terrain, I don't think it's very wise.

Table 7-1. Mountain flying tips

- Plan en route stop at one of the foothill airports prior to entering mountainous terrain.
- Consult a local accident-prevention counselor for advice on routing, weather, and other tips.
- Check the weather over your entire route. Do not attempt the flight if winds aloft near the mountain tops exceed 40 percent of the aircraft's stall speed. If weather is marginal, delay the trip.
- Plan trips during the early morning or late afternoon hours.

Table 7-1. Continued.

- Use current charts, preferably sectionals or state air navigation charts. Radio navigation may prove difficult due to high terrain.
- Route your trips over valleys wherever possible.
- Learn as much as possible about your intended destination airport.
- Carry enough fuel to make your trip with ample reserve.
- Know your aircraft's performance and limitations.
- Make proper corrections for pressure and temperature effects on takeoff and rate of climb.
- Check weight and balance of loaded aircraft before takeoff.
- Your normal horizon is near the base of the mountains.
- Beware of rapidly rising terrain and deadends in valleys and canyons.
- Downdrafts and turbulence occur on the lee side of mountains and ridges.
- Approach a ridge at an angle so you can turn away if you encounter a downdraft.
- Maintain flying speed in downdrafts.
- Carry survival equipment. Even summer nights are cold in the higher altitudes.
- Be prepared for downdrafts and turbulence on final approach.
- Use power on approaches.
- FILE A FLIGHT PLAN.

(Compiled and published by the FAA.)

POWER LINE AVOIDANCE

Towers, poles, and their electrical lines might not always be readily visible. Most are *not* required to be marked under the FAA criteria that determine what is considered to be an obstruction to air navigation. Also, under some conditions, such as sun glare or haze, it can be difficult to see the lines running between the support structures (Fig. 7-9). Taking two simple steps can greatly reduce the chances of accidently contacting electrical facilities while you're flying.

First, take time for safety planning prior to the flight. Check the aeronautical charts for obstructions when you plan the route. Certain electrical lines are charted because of their height. Also, airport directories carry warnings of power lines located close to runways.

Second, observe the minimum altitude requirements while airborne, especially the 1,000-foot minimum over populated areas. It's a good idea to stay above 1,000 feet when flying over lakes, rivers, or canyons to avoid any power line crossings.

Table 7-2 is a good summation of power line avoidance tips.

Fig. 7-9. *Towers holding power lines are usually not too difficult to see, but the power line cables can be all but invisible under most conditions. Notice how the cables disappear in the lower left corner of the photograph.*

Table 7-2. Power line avoidance tips

- Before you take off, check the airport directory for warnings about power lines at your destination.
- Check your route and be familiar with marked obstructions.
- Always observe altitude minima.
- Do not allow adverse weather conditions to force you to fly too close to the ground. Check aviation weather forecasts to make sure cloud heights provide an adequate ceiling for safe visual flying.
- Remember that sun glare can make power lines nearly invisible.
- Maintain a safe altitude over rivers, lakes, and other waterways.

Table 7-2. Contnued.

- When visibility is poor, increase your altitude above minima or fly instrument flight rules.
- Be aware that power lines and towers are marked only near airports and at certain water crossings.
- When you're using private airports, check for nearby power lines. Call ahead for information about any obstruction at any private field you're planning to use.
- When flying through a gap or over a mountain ridge, watch for winds and turbulence that could force you into a power line crossing.

(Compiled and published by the Pennsylvania Power & Light Company in conjunction with the FAA).

OVERWATER FLYING

"I slip out through the big, half-open door, and stare at the glowing mist above Garden City. That means a low ceiling and poor visibility—street lights thrown back and forth between wet earth and cloud. The ground is muddy and soft. Conditions certainly aren't what one would choose for the start of a record-breaking flight. But the message from Dr. Kimball says that the fog is lifting at most reporting stations between New York and Newfoundland. A high-pressure area is moving in over the entire North Atlantic. The only storms listed are local ones, along the coast of Europe."

Charles A. Lindbergh, *The Spirit of St. Louis*

If you fly only in the 48 United States, you might never really need to fly over water for any extended period of time, as Charles Lindbergh did on his solo flight across the Atlantic. On the other hand, if you've ever thought of flying to the Bahamas or one of the islands in the Caribbean, you certainly will find yourself over the sea for quite awhile. Sometimes even in the continental United States, the most logical route is along the coast or over a large lake. Whatever the reason for flying over water, additional precautions should be taken (Fig. 7-10).

First of all, filing a flight plan is always a good idea for extended over water flights. If your route takes you into the coastal or domestic ADIZ/DEWIZ, a defense VFR (DVFR) flight plan is required for security reasons. If you do have to ditch, a filed flight plan is your best insurance that someone will note your absence and send out an alert. Search and rescue procedures are started one-half hour after your estimated time of arrival if your flight plan isn't closed or canceled. Because the probability of survival after a ditching decreases rapidly over time, the sooner rescue services start looking for you, the greater are your chances of surviving.

Fig. 7-10. *Incorrect fuel calculations take on new meaning during long flights over water.*

Navigation and communication requirements

Navigating over water is obviously different from navigating overland. Map reading is impossible, except when you have land in sight, and this is sometimes difficult because long stretches of coastline might appear very similar with few identifiable landmarks. Making the situation even more confusing, wave action, storms, and normal tides can significantly change the shape of coastlines from what is depicted on a chart; whole islands can appear and disappear.

Finding usable navaids is sometimes a problem. VORs often can be used to great distances, especially if you fly high, but are still subject to the limitations of line of sight. If the VOR you want to track is located in a valley, you'll either have to be very close or very high to receive it. An NDB might be the only navaid you can receive at distances over 100 miles from land; although not always precise, a bearing from an NDB is certainly better than nothing at all.

Strangely, the FARs do not specify any particular requirements for single-engine aircraft operating under Part 91 when flying overwater. Perhaps the Feds don't expect anyone to do it. Paragraph 91.511, "Radio equipment for over water operations," is in a subsection of the FARs that applies only to large and turbine-powered multiengine airplanes. This can, at least, be used as a guide if you plan to fly overwater often.

Basically, if you plan to fly over water more than 30 minutes or 100 nautical miles from the nearest shore, you should have at least two transmitters, two microphones, two headsets (or one headset and one speaker), and two independent receivers that are operable and "appropriate to the facilities to be used and able to transmit to, and receive from, any place on the route, at least one surface facility." With respect to navigation equipment, the aircraft should have "two independent electronic navigation units capable of providing the pilot with the information necessary to navigate the airplane within the airspace assigned by air traffic control." Again, these requirements are in a subsection of the FARs that applies to large and turbine-powered multiengine airplanes.

With the proliferation of loran and GPS receivers, it's much easier than it used to be to have good electronic navigation equipment in single-engine airplanes. Loran was originally implemented for marine navigation; therefore, it provides better coverage overwater than it does overland in many countries. GPS is excellent for navigating overland and water as long as the receiver is able to keep enough satellites in view. VLF/OMEGA, which uses land-based stations developed for marine navigation, is an excellent area navigation system; however, VLF/OMEGA receivers are much more expensive than many loran and GPS receivers. Top-of-the-line navigation systems include provisions for combinations of loran, GPS, and VLF/OMEGA.

Your backup overwater navigation method is dead reckoning. To be accurate, dead reckoning requires good route planning, careful heading and airspeed control, a halfway decent idea of the wind speed and direction, and a well-maintained flight log. Like anywhere else, winds offshore can be very changeable sometimes, particularly in conjunction with frontal weather, but very often the wind will be very steady over large stretches of water. With a good weather briefing prior to departure, you will be able to compute drift angle, heading, ground speed, and time en route with a very good degree of accuracy and confidence.

Many pilots will only fly over water in multiengine aircraft. Others will fly single-engine aircraft as long as certain conditions are met. A good rule for short overwater flights in single-engine aircraft is to fly high enough to be within gliding distance of land (Fig. 7-11). Obviously, this is not always possible; you might not be able to fly high enough because of weather or the operating capability of the airplane; additionally, the land within gliding distance might be steep, rocky, or otherwise unsuitable for a power-off landing. Carry emergency survival equipment to mitigate the dangers of ditching.

Overwater survival equipment

The FARs say little about survival equipment requirements for overwater flights in single-engine aircraft. Only in a portion of FAR 91.205 is anything mentioned, and even that concerns for-hire aircraft:

"For VFR flight during the day, the following instruments and equipment are required: . . . If the aircraft is operated for hire over water and beyond power-off gliding distance from shore, approved flotation gear readily available to each occupant and at least one pyrotechnic signaling device. As used in this section, 'shore' means that area of land above the high water mark and excludes land areas which are intermittently under water."

FAR 91.509, "Survival equipment for overwater operations," also applies only to large and turbine-powered multiengine airplanes. The regulation gives some clues about the other things you should consider to take.

Piper Aircraft Corporation

Fig. 7-11. *If you can't fly high enough to be within gliding distance of land, you should carry overwater survival equipment.*

For flights that take a multiengine airplane no more than 50 nautical miles from the nearest shoreline, FAR 91.509 requires that a life preserver or other suitable flotation device (for example, a floatable seat cushion) be carried onboard for each occupant.

For overwater flights that last more than 30 minutes, or are 100 nautical miles or more from the nearest shore line, large and turbine-powered multiengine airplanes are required to carry the following: a life preserver with an approved locator light for each occupant; enough life rafts to accommodate all occupants; at least one pyrotechnic signalling device for each life raft; an independent, floatable, water-resistant emergency radio signaling device; a lifeline; and a survival kit attached to each life raft.

Swimming 50 miles to the nearest shore is a long way to go, even with a life preserver; 25, 10, or 3 miles might be too far. When was the last time you swam a mile? If you ever have, it was probably in a pool or small lake. Swimming in the open sea or a large lake is a lot different.

For anyone contemplating a long overwater flight, life vests *and* a life raft are absolute musts, if you want any chance of surviving a ditching. "Long overwater flight" can be defined as any flight that takes you farther from shore than the sum of the airplane's gliding distance and the distance you are able to swim (Fig. 7-12).

Even if you are an excellent swimmer and in superb physical shape, don't forget about hypothermia. Your body will lose heat to the sea, regardless of the water temperature, if you don't get out of the water and into a life raft.

It's a matter of basic physics. The thermal conductivity of water is 240 times greater than that of still air. This means that water or wet clothing can extract heat from your body up to 240 times faster than still air or dry clothing.

Robert Hoffman/The Bear Valley Voice

Fig. 7-12. *You don't have to fly far offshore to end up in the water. The pilot of this Beechcraft Bonanza had to ditch in a lake after the engine lost power shortly after takeoff. Fortunately, he was uninjured, although he had to be treated for hypothermia after wading to shore. Because the lake was only three feet deep, the airplane didn't sink.*

The colder the water, the faster you lose body heat. In 40°F water, your effective time is not much more than 30 minutes. After two hours, you have a 99 percent chance of being dead. Even in water warm enough to comfortably swim in, the average person can't last much longer than 12 to 15 hours before hypothermia sets in. The life raft is a crucial part of survival after ditching.

If you have to make a long flight over cold water, seriously consider taking or wearing a survival suit. Wet clothing is almost as bad as being immersed in water. When clothing gets wet, it no longer provides an insulating layer of warm air next to the skin. The wet fibers rapidly conduct heat away from the body to be dissipated into the outer environment. Wet clothing is like a wick. If a cold wind is blowing, this "water chill" will dissipate heat much more rapidly than the body can produce it.

Survival suits. Survival suits vary in price and sophistication. Relatively inexpensive "rubber balloon" suits are designed to be carried in a small bag and donned just prior to or after ditching. Expensive full-fledged flight suits have fire-resistant Nomex outer layers and breathable, waterproof inner linings. Realize, however, that even the best water survival suits cannot prevent heat loss if a person is immersed in very cold water over an extended period of time. The only sure way to protect against hypothermia is to use both a survival suit and a life raft.

Another important reason for using a life raft is that it is much easier for rescue aircraft to spot a life raft than a person floating in the water with only a life vest. If there are any waves at all, it's just about impossible to see someone in a life preserver unless the search aircraft happens to fly directly overhead. Believe me, I've done it.

Which brings up the question of signaling devices. Probably 99 times out of 100 a person lost at sea will see or hear a ship, boat, or aircraft long before the crew in that boat or aircraft sees someone in the water, even if they are deployed expressly for the search. Rescuers will find you much faster if you help them by signaling.

Electronic signaling devices. The best way to bring rescue craft to your location, of course, is with an emergency locator transmitter (ELT). A floating, water-activated ELT attached to the life raft is easy to activate and will bring rescue aircraft to your position, rather than the airplane's. Wind and waves will separate the life raft from a floating airplane even under relatively calm conditions.

Voice communication with rescue aircraft will speed up your rescue. If it's waterproof, a hand-held radio will fulfill this task nicely. If it's not waterproof, it probably won't last very long in a life raft. The alternative is an emergency radio. Unfortunately, these are expensive, but an emergency radio might well be worth the cost and peace of mind if you plan to do extensive overwater flying. It can be tricky to operate; therefore, be sure to read and understand the instructions before you really need to use it. Don't forget to buy an extra set of batteries, and keep all of them charged.

Less expensive signaling devices include mirrors, flares, and sea dye. Some life rafts come with survival kits that have all three; if not, buy whatever is missing because all three types are useful in different situations.

Mirrors. Mirrors reflect sunlight toward an airplane or boat. They are very effective in most daylight conditions, unless the sun is totally obscured. In certain light conditions, a mirror might be more effective than any other signaling device, even a radio. Good signal mirrors are made of rust-free metal and feature a sighting hole to aim the reflection. Any shiny material can be used to reflect the sun, even a piece of broken glass or a lamp reflector. For the money though, a signal mirror is probably the best signaling device you can buy. The mirror never wears out, has no parts to break, always works when the sun shines, and is very effective.

Flares. Flares come in two basic types: day or night. Each type can be used anytime. Day flares put out more smoke than light, whereas night flares put out more light than smoke. If you've used up all your night flares, popping a day flare at night is better than doing nothing at all because a day flare does produce some light. Similarly, night flares can be seen during the day because they do produce some smoke. In fact, on dull, murky, overcast days, the light generated by a night flare is more visible than the smoke from a day flare.

You can buy hand-held flares and projectile flares. Again, both are worth taking along. Projectile flares obviously permit you to advertise your position to a greater distance; hand-held flares pinpoint your exact location. Smoke from a hand-held flare also shows the wind direction, which will help searchers determine the best direction to make their approach; watercraft and aircraft will need to compensate for a drifting life raft that is pushed by the winds.

Sea dye. Sea dye might not seem like a very effective signaling device. The dye will make a big "spot" in the water, much bigger than either a life preserver or a life raft. That extra-large spot can really help searching aircraft find your position. Depending on sea conditions, sea dye might last much longer than a flare. You have no way of knowing how long the dye will remain at your position until you put it in the water. Water currents, waves, and winds might quickly separate you from the dye. Wait until you see an aircraft approaching before using it.

The dye has even proven to be a shark repellent, in some instances. The danger of shark attack is extremely small, but a person in a life raft will be much less attractive to a shark than someone treading water. According to shark experts, the noise a swimmer makes sounds like an injured fish, which attracts the shark. If you're bleeding and treading water, watch out.

Remember, sharks, like people, prefer warm water; warmer water increases chances of sharks in the area. On the other hand, cold water hastens hypothermia. In a life vest, you can't win, no matter what the water temperature. Buy a life raft and survival kit. Explore the possibility of renting a raft and kit if you don't plan on repeated overwater trips.

Survival kit. The contents of a survival kit will depend a great deal on where you'll be flying. Whole books have been written on the subject of survival, one of the best is the U.S. Air Force's AFM 64-5 survival manual. The manual might be available in your local library. Study this manual, and decide what you want to carry. You might find that a ready-made survival kit is sufficient for your needs. You might want to supplement a ready-made kit with additional items, or create a personally customized kit from scratch. (*See* appendix A.)

A water survival kit will differ somewhat from a land survival kit. Your main concerns after a ditching will be to stay warm and to signal search aircraft and vessels. Fresh water is an important secondary concern, but you can survive a few days without it. If you filed a flight plan, you'll probably be found and picked up within 24 hours. Food is only a minor concern after a ditching because you'll probably become seasick anyway, unless you happen to be a particularly hardy sailor.

The possibility of administering first aid will be limited because everything will be wet. Plan on being uncomfortable. If you have the basics—a life raft, a protective covering, some signaling devices, and perhaps a bottle or two of water—you'll probably survive.

Ditching

Ditching an airplane, any airplane, under any conditions is going to be tough. Think about it this way. How long did it take you to learn how to make a decent landing on a normal runway? How long did it take you to learn how to land on a short runway? A gravel runway? With a crosswind? How often do you grease a landing? How often do you practice power-off landings?

How many power-off landings have you made on water with a landplane? Probably none. It's not a practice maneuver that flight schools normally include in their training curriculum.

If you ever have to ditch, accept the fact that you're going to be a student pilot again. Not only that, you're going to be a test pilot, too, because no manufacturer sends test pilots out to land on water to determine the ditching characteristics of their landplanes. What happens to you after you hit the water is guesswork. Except for two things:

- The airplane will probably capsize.
- The airplane will probably sink.

Bank on it. Bet your life on it. Prepare for it. *Plan* on being inverted after a ditching. If you do happen to beat the odds and find yourself sitting right-side-up with the airplane still floating on the sea, congratulate yourself on your extreme good luck only after exiting the aircraft, accounting for all souls, deploying the life raft, and activating the ELT. Never stretch your luck by believing that the airplane will stay afloat at your convenience.

The *Airman's Information Manual* gives a good deal of advice about how to ditch an airplane. If you have a photographic memory, you might recall some of it in the stress of the moment when the engine quits over water 100 miles from the nearest land. If you're a normal person, you probably won't remember much at all.

To counter brainlock at such a critical moment, use some time before you take off to prepare a ditching checklist for your airplane, and post it where you can find it quickly. Use the information in the AIM and the following suggestions when you make your checklist.

In general, prepare as you would for a controlled crash on land. Attain the slowest speed and rate of descent that permit safe handling, and turn into the wind. Wind direction is a "best guess" thing. If you have a loran, GPS, or other area navigation device that determines wind direction, use that. If you don't have such a device, line up with the wind streaks on the sea. It's easy to misinterpret wind streaks by 180°, so always maintain an awareness of the wind direction when flying over water.

If you have retractable gear, leave it up. If you lower it, it will either shear off when it hits the water or will flip the airplane upside-down sooner than with the gear up.

Make a quick Mayday call on whatever frequency you're monitoring. Don't bother to switch to 121.5 for your first distress call. Your best chance of someone hearing you is on the frequency you last communicated on. Repeat "Mayday" three times, give your callsign, your position, and say you're ditching. If you have the time (i.e., altitude) to make two calls and no one has acknowledged your first call, then try on 121.5.

Put on your life vest, if you have time, and tighten your seat belt and shoulder harness.

As you descend lower, you'll be able to see the direction of the waves and the wind much easier. The AIM goes into much detail about oceanographic terminology: swell face, fetch, chop, and the like. AIM warning: "It can be extremely dangerous to land into the wind without regard to sea conditions." Unfortunately, taking regard for sea conditions is not always easy.

To the landlubber, a wave is a wave is a wave, but strictly speaking, a *wave* is "the condition of the surface caused by local winds." A wave that is left over from a storm or is caused by a distant disturbance is called a *swell*. More than one swell condition can exist in one patch of water. It's not uncommon for the wind-created waves to be going in one direction, the primary swells in another direction, and the secondary swells in a third. Add to this the fact that there are riverlike currents flowing in the ocean, and you can begin to get an idea how complex the situation can become.

The main point to remember is to avoid hitting the face of a swell, which is the side of the swell toward the observer. In other words, don't land going into a wall of water.

This is much easier said than done. With the wind going in one direction and the swell going in another, you'll be making a crosswind landing to a surface that is moving both horizontally and vertically. As you descend just above the water, ground effect will alternately take hold and let go as the waves and swells roll beneath you. You can try, as the AIM recommends, to land on a crest or on the backside of a swell. Chances are, even if you manage to do this, you'll meet the face of a swell sooner or later. The nose of the airplane will plow in, the airplane will do a somersault, and you'll find yourself with a fish-eye view of the ocean.

Escape becomes the main concern. Unless the force of the ditching has broken the windshield, you'll be in a pocket of air inside the cabin. Release your seat belt, and turn yourself right-side up. Grab your life raft and survival kit, and get ready to go swimming. As necessary, account for and attend to any passengers.

The force of the water will make it impossible to open the doors until the pressure inside and outside the cabin are equal. Water leaking in might equalize the pressure for you, but you can speed up the process by opening or breaking a side window.

Brace yourself so that you're ready to kick the door open, unlatch the door (or pull the emergency release), take a deep breath, and open the side window. Expect to be shocked by the coldness of the water, unless you're in the tropics or wearing a good survival suit.

As soon as enough water has entered the cabin to equalize the pressure, kick the door open, swim through the opening, inflate your life vest, and float to the surface. Never inflate your life vest inside the cabin because it will make it harder for you to get out and might become punctured in the process.

Most life rafts will float even before they're inflated, but it's a good idea to check this before you have to ditch; otherwise you might be pulled downward as you try to swim upward. If your raft does sink before it's inflated, pull the inflation lanyard after you exit the aircraft (never before, unless you're floating upright on the surface in a calm wind), and hold on tight to the cord. In strong winds, an inflated life raft will have a tendency to fly like a kite. If you don't hold onto the lanyard, the raft will scoot away faster than you'll ever be able to swim after it. Climb into the raft, make sure the ELT is transmitting, and get your signaling devices ready.

Doing this maneuver right the first time without practice will take a lot of luck. Pilots who frequently fly over water and want to improve their odds of surviving a ditching should consider attending a school that specializes in underwater egress training. The course doesn't take longer than a day or two, is not overly expensive, and is the closest thing to a real ditching that most of us will ever want to experience. Few people enjoy training in a "dunker," but if you ever do have to make a real ditching, a couple of practice ditchings in a swimming pool could well make the difference between being a survivor or becoming a statistic.

Overwater weather

Water is the fuel of weather. With so much "weather fuel" under you as you fly over the water, you'd expect to find all sorts of weather. You'd be right. From relatively calm advection fog to the violence of hurricanes and typhoons, overwater weather is anything but boring.

Unlike mountain weather's unpredictability, overwater weather is much more uniform; you don't encounter the numerous microsystems as you do in the mountains. This doesn't mean that you won't encounter changing weather over water. Global air circulation patterns caused by equatorial heating, polar cooling, and the earth's rotation contribute to the mixing of warm and cold air in the area between 30° and 60° latitude.

This belt of prevailing westerly wind in the Northern Hemisphere is characterized by migrating storms, frontal waves, and alternating high- and low-pressure systems. The weather does change, often rapidly; however, the changes occur over a relatively wide area and are usually not difficult to forecast.

Because of currents, uniformity, wave action, and other factors, the temperature of large bodies of water varies slowly. Whereas the air temperature might rise or drop 10s of degrees in a matter of minutes, water temperature needs days and weeks to change only a degree or two. The layer of air from the surface of the water to a few hundred feet above it will be influenced by the water temperature; therefore, the layer of air at the surface is frequently warmer or cooler than the surrounding air mass.

For example, helicopter pilots flying to and from oil platforms in the North Sea have learned that icing conditions rarely exist below 500 feet, even in the winter, due to the warming influence of the water. Low-level temperature inversions are also common over water.

In the Caribbean, the weather tends to be much more benign than in the North Sea. The influence of the ocean on the atmosphere above it causes the air temperature and dew point to converge at approximately 2,000 feet, creating cloud bases at this level. Lower ceilings are rare; in fact, if the ceiling is below 1,000 feet, there's a good chance a hurricane is brewing.

Warm moist air flowing over colder ground or water creates *advection fog*, or *sea fog*, as it's called over water. Unlike radiation fog, which forms when the temperature-dew point spread is small and there's little wind, advection fog can be unaffected by strong winds; I've seen sea fog when the surface wind was 40 knots. Advection fog is also much more tenacious than radiation fog. Radiation fog will usually burn off as the sun rises higher in the sky; advection fog can hold on for days with "zero-zero" intensity. Advection fog can also be carried over land, as residents of coastal cities well know; sea fog from the Gulf of Mexico is common over the southeastern United States, for example.

Advection and radiation fog sometimes work together. When conditions are just right, sea fog might hug the coastline all day, with clear skies over land; as the air temperature decreases toward evening, radiation fog forms over the land surfaces and eventually combines with the sea fog to form one solid cloud mass.

Land and sea breezes. Coastal residents are also familiar with sea and land breezes (Fig. 7-13). Because land and water surfaces absorb different amounts of heat, convection over these two kinds of surfaces occurs at different rates and causes wind. Land warms and cools faster than water. As land heats up and convection takes place during the day, the air pressure over the land becomes lower than that over water. A pressure gradient forms, causing the higher pressure air over the water to flow toward the lower pressure air over land. The resulting wind is called a *sea breeze* because it moves from the sea to the land.

In the evening, the opposite takes place. The land loses heat rapidly due to radiation. Water loses heat at a much slower rate. A pressure gradient forms again, this time with the cooler, high-pressure air that is over the land moving toward the relatively warmer, low-pressure air that is over the water. This is a *land breeze*.

Fronts. Frontal weather over water is much the same as it is overland. If anything, it might be a little less complicated because it doesn't have the effects of orographic turbulence and uneven surface heating to contend with. Don't take this to mean that offshore weather doesn't get intense or complicated. I've seen as many as five fronts pinwheeling around a single low-pressure system in the North Sea. Low ceilings and visibilities, high winds, heavy rain, icing, and turbulence are not uncommon. On the other hand, thunderstorms over water, except those associated with tropical storms, usually do not grow as large as those overland, although they should still be avoided.

Storms. Tropical cyclones are the deadliest storms on Earth. These storms are born in the late-summer environment of the tropics when rapid evaporation of ocean water caused by the sun transfers heat rapidly into a developing cyclone (low pressure). The evaporation rate controls the formation of the storm; at ocean temperatures below 76°F, cyclones cannot develop.

During the Atlantic hurricane season, about two seedling storms a week drift westward across the Atlantic at about 20 mph. At first, these storms produce only innocuous rain showers, but when conditions are favorable, vigorous thunderstorms and high-level winds create more severe thunderstorms. Eventually a tropical storm devel-

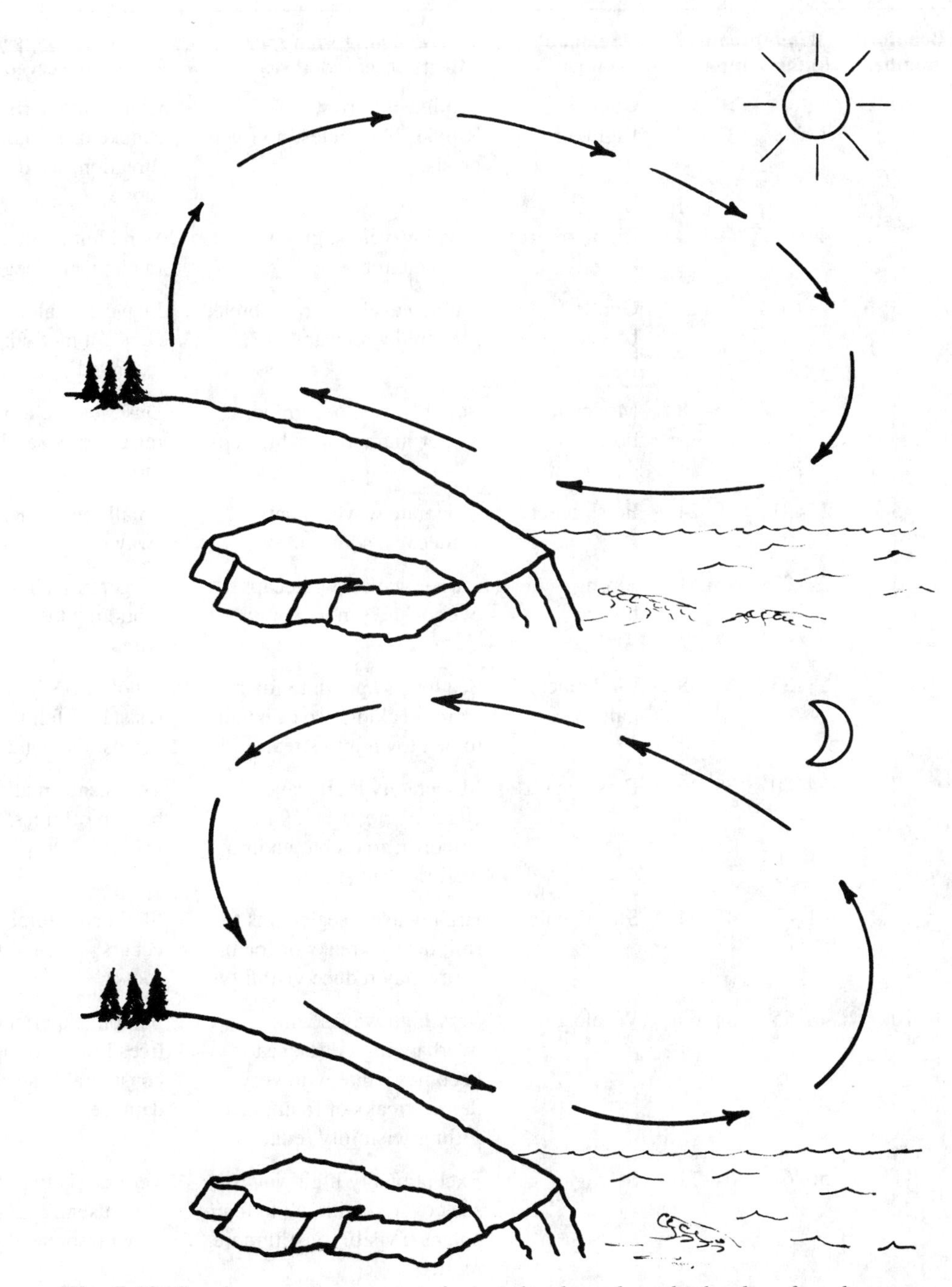

Fig. 7-13. *Sea breezes (top) occur during the day when the land surface heats up faster than the water and the cool air over the water flows toward the land. Land breezes (bottom) occur at night because the land loses heat faster than water, cooling the air over its surface. This cool air then flows outward toward the sea.*

Table 7-3. Beaufort wind scale

Beaufort	*Wind speed*		Seaman's	*Estimating wind speed*	*Estimating wind speed*
number	**knots**	**mph**	**term**	**Effects observed at sea**	**Effects observed on land**
0	>1	>1	Calm	Sea like a mirror.	Calm; smoke rises vertically.
1	1–3	1–3	Light air	Ripples like scales; no foam crests.	Smoke drift indicates wind direction; wind vanes do not move.
2	4–6	4–7	Light breeze	Small wavelets; glassy crests; not breaking.	Wind felt on face; leaves rustle; vanes begin to move.
3	7–10	8–12	Gentle breeze	Large wavelets; crests break; scattered whitecaps.	Leaves; small twigs in constant motion; light flags extended.
4	11–16	13–18	Moderate breeze	Small waves, becoming longer; numerous whitecaps.	Dust, leaves, loose paper raised up; small branches move.
5	17–21	19–24	Fresh breeze	Moderate waves; many whitecaps; some spray.	Small trees in leaf begin to sway.
6	22–27	25–31	Strong breeze	Larger waves; whitecaps everywhere; more spray.	Larger branches in trees in motion; whistling heard in wires.
7	28–33	32–38	Moderate gale	Sea heaps up; white foam from breaking waves begins to be blown into streaks.	Whole trees in motion; resistance felt in walking against the wind.
8	34–40	39–46	Fresh gale	Moderately high waves; edges of crests break into spindrift; foam blown into well-marked streaks.	Twigs and small branches broken off trees; difficult to walk.
9	41–47	47–54	Strong gale	Hight waves; sea begins to roll; dense streaks of foam; spray may reduce visibility.	Slight structural damage occurs; slate blown from roofs.
10	48–55	55–63	Whole gale	Very high waves with overhanging crests; sea becomes white with very dense streaks of foam; heavy rolling; visibility reduced.	Seldom experienced on land; trees broken or uprooted; considerable structural damage.
11	56–63	64–72	Storm	Exceptionally high waves; sea covered with white foam patches; visibility still more reduced.	Very rarely experienced on land; usually accompanied by widespread damage.
12	>64	>73	Hurricane	Air filled with foam; sea completely white with driving spray; visibility greatly reduced.	Violence and destruction.

ops. Fortunately, out of the 100 or so seedling storms tracked each year, only about nine develop into gale-force tropical storms or full-fledged hurricanes.

Thunderstorms associated with tropical storms often rise up to over 50,000 feet. The winds are strongest at the surface, but have been measured over 100 knots as high as 18,000 feet. The speed of a storm's vertical winds—the updrafts and downdrafts—can exceed 30 knots. Turbulence increases toward the eye of the storm, becoming most violent in the wall cloud surrounding the eye. As if these hazards weren't enough, the intense low pressure of a cyclonic storm can cause erroneous altitude readings from pressure altimeters.

Ever wonder why flying magazines have so few articles about how to fly in hurricanes? The reason is simple. You shouldn't do it. Nobody should do it, except for the pros working for the National Weather Service, military, or NASA. They fly four-engine Lockheed P-3 Orions and C-130 Hercules. If you fly anything smaller than these 145,000-pound and 175,000-pound turboprop airplanes, don't even think about flying in a hurricane.

Sea spray

One little-known overwater hazard is sea spray. Sea spray occurs when the wind lifts the spray off breaking waves and carries it upward. You can encounter sea spray with winds as low as 15 knots when flying low, but usually you can avoid it by flying above 500 feet. Generally, the stronger the wind, the higher it will lift the spray. I've encountered sea spray as high as 1,500 feet when the wind was about 50 knots. You'll know you're getting sea spray when the windshield begins to look dirty. Usually, the mist is so light that you don't notice any moisture because it dries as soon as it contacts the aircraft. After awhile, you see that the windows aren't as clear as they were when you took off.

Sea spray is not a problem over fresh water, but over salt water it can have a detrimental effect on engines and lifting surfaces. The sea spray deposits a layer of salt on the wings of an airplane and rotor blades of a helicopter, which reduces the lifting capabilities, similar to icing. Salt deposits also reduce the effectiveness of engines. Turbine engines are more susceptible than reciprocating engines because the salt deposits reduce the efficiency of compressor blades. Loss of power and airspeed is a consequence in turbine-powered aircraft. In recip-powered airplanes, the biggest problem is usually no more than dirty windshields. Flying through a small rain shower will often remedy the problem in flight. The only other solution is to wash the aircraft with fresh water after you land.

Fuel

If an overwater flight is not far from land, perhaps along a coastline or over a large lake, then your fuel requirements are not going to be much different than overland flying; however, if you plan to fly a long distance over water, perhaps from the mainland to an island, you definitely don't want to skimp on fuel (Fig. 7-14). In fact, if the distance is great enough, you might even need to add an auxiliary fuel tank. International regulations require a minimum of a 3-hour reserve for each leg of transatlantic flights. Professional ferry pilots try to carry a 50- to 100-percent reserve whenever they can.

Piper Aircraft Corporation

Fig. 7-14. *Don't skimp on fuel if you plan to fly a long distance over water. Professional ferry pilots prefer to take 50–100 percent more fuel than would be necessary for a flight overland.*

Even if you plan to fly VFR, carry enough fuel to reach your destination, proceed to an alternate airport that has good weather forecasted, and be able to land with at least an hour's worth of fuel in the tanks. Be very conservative with your calculations when planning fuel. Figure on the strongest winds forecast. Round up an odd amount up to the next higher amount. No one wants to run out of fuel overland, but running out of fuel over water is worse.

While I was in Iceland, one of our helicopter crews picked up a pilot who was trying to make it across the Atlantic. He ran out of fuel on the leg from Greenland to Iceland, only 5 miles from Keflavik Airport. If he had run out of fuel over land, he might have been able to deadstick to a road or field with subsequently little damage to the airplane. Unfortunately, 5 miles out of Iceland's Keflavik Airport is over water. He survived the ditching because the search and rescue crew was able to pick him up only minutes after the airplane hit the cold North Atlantic water.

Two calculations are useful to the overwater pilot: point of no return and point of equal time. Determining these two points during preflight planning will provide useful information to establish fuel requirements for a particular flight.

Point of no return (PNR)

This ominous-sounding phrase has a much less dramatic meaning than most people think. The *point of no return* is the point beyond which an aircraft will not have enough fuel to return to its point of origin. In other words, after passing the point of no return, a pilot is committed to continue toward the destination. As a matter of interest, most offshore helicopter operators require their pilots to carry "return-to-land" fuel when flying to offshore destinations. This means that the point of no return must always be at or past the oil rig or platform they are flying to.

If you were planning to fly to an island, and the closest usable alternate was your departure airport, you would want your point of no return to be past the island. This would ensure that you had enough fuel to fly to the island and then return to your point of origin without refueling, a useful option if weather makes it impossible to land at your island destination.

To find the point of no return between two points, you need to calculate the ground speed in both directions of flight, plus calculate the total endurance. The following equation applies:

$$\frac{GS(ab) + GS(ba)}{GS(ba)} = \frac{TE - \text{reserve}}{\text{Minutes to PNR}}$$

where
GS(ab) is ground speed from point A and point B,
GS(ba) is ground speed from point B and point A, and
TE is total endurance in minutes.
Reserve is fuel reserve in minutes.

For an example, let's say that the distance between A and B is 175 nautical miles. Ground speed from A to B is 130 knots and from B to A is 105 knots. Total endurance is 3:40. Reserve is 30 minutes. Therefore,

$$GS(ab) + GS(ba) = 130 + 105 = 235 \text{ knots}$$
$$TE - \text{reserve} = 3{:}40 - 0{:}30 = 190 \text{ minutes}$$

Step 1: On a flight computer, set 105 on the inner scale under 235 on the outer scale. This establishes the ratio GS(ab)+GS(ba) over GS(ba).

Step 2: Now find 190 on the outer scale. Below it on the inner scale is 85. This is the time in minutes it will take to fly to the PNR on the track A to B. Because the estimated time en route from A to B is 1 hour, 21 minutes (175 nm at 130 knots), and the time to the PNR is 85 minutes, or 1 hour, 25 minutes, we know that we can fly all the way to B and still have enough fuel, plus a 30-minute reserve, to return to A.

Step 3: To find the estimated time of arrival at the PNR, add 85 minutes to the time the aircraft passed over point A.

The calculations can be checked in the following manner. We have found that the PNR is located 1 hour, 25 minutes, from point A. At 130 knots, we will fly 184 nautical miles in 1 hour, 25 minutes. If we fly all the way to this point (which would take us 9 nautical miles past point B), we should have just enough fuel to fly back to point A and still have our original 30-minute reserve.

Flying 184 nm at 105 knots gives 1:45 from point B to point A. Flying 1:45 (B to A) plus 1:25 (A to B) equals 3:10 (total flight time). Flying 3:10 plus 0:30 (reserve) equals 3:40 (total endurance). The PNR calculations are correct.

Point of equal time (PET)

The point of equal time is that point between two points from which it will take the same time to fly to either point. If there is no wind, the point of equal time is precisely halfway between the two points. If there is a tailwind between the point of origin and the destination, the PET will be closer to the point of origin. If there is a headwind between the point of origin and the destination, the PET will be closer to the destination.

The PET is useful if you encounter an emergency and you want to land as soon as possible. Normally, you don't want to ditch unless you really have to, so the goal is to find the shortest route, in time, to a suitable landing site. Before reaching the PET, returning to your point of origin will get you on the ground sooner than if you flew to your destination. After passing the PET, continuing to your destination is the better course of action, all other things being equal.

To find PET, we use the same ratio on the left side of the equation that we used when figuring the PNR.

$$\frac{\text{GS(ab)} + \text{GS(ba)}}{\text{GS(ba)}} = \frac{\text{Total distance one way}}{\text{Distance to PET}}$$

To illustrate how to find PET, let's use the same conditions as in the PNR example.

Step 1: Set 130 on the inner scale under 235 on the outer scale.

Step 2: Find the total distance one way (175 in our example) on the outer scale. Under it, read 78. This is a distance figure. It means that from point A to the point of equal time is 78 nm.

Step 3: Now find the time to PET by setting the black triangle under the ground speed from A to B, 130 knots. Find 78 on the outer scale and under it read 36. From point A to PET is 36 minutes.

In checking our calculations, we know that from the point of equal time it should take just as much time to fly to point A as it takes to fly to point B.

From the PET to point A, the distance is 97 nm (175 nm – 78 nm). At 130 knots, this will take 44.5 minutes. The distance from the PET to point B is 78 nm. At 105 knots, this will take 44.5 minutes; therefore, the PET calculation is correct.

HOW TO AVOID BIRDS

> "A mile from shore a fishing boat chummed the water, and the word for Breakfast Flock flashed through the air, till a crowd of a thousand seagulls came to dodge and fight for bits of food. It was another busy day beginning."
>
> **Richard Bach, *Jonathan Livingston Seagull***

Pilots don't usually think of birds when they think of adverse conditions (Fig. 7-15). The fact is, however, a single bird can do more damage to an airplane in a second than an entire hour of flight in moderate turbulence or light icing. Like the hazards of thunderstorms, the best way to cope with the hazards of bird strikes is to avoid them.

Although birds are a hazard for all aircraft, the faster you fly, the greater the hazard because a bird will impact harder at the faster speed. The equation for force says it

Fig. 7-15. *Birds aren't usually a hazard to aircraft on the ground, except when they build nests in the engine compartment. Birds in flight can be more hazardous than icing or turbulence.*

all: F = ma (force equals mass times acceleration). The mass of a bird is not much compared to an airplane. The bird might not be traveling very fast, but the airplane will be, and the resultant rate of closure between the two is therefore high. That acceleration times the mass (weight) of the bird creates the force that can damage an airplane.

I've never had a bird come through a windshield. I hope that it never happens, but I can imagine what it would be like. Even the slowest airplanes travel at least as fast as a car on a highway. Several times a car in front of me has kicked up a small rock and sent it on a collision course for my car's windshield. It usually happens so fast that there's no time to react, except to be shaken up afterward when considering the consequences. Fortunately, the safety glass has always held, and dime-size pock marks on the windshield are the worst damage that a car has ever received.

The rocks that have hit my cars' windshields have been no bigger than pebbles. Consider how much greater the force would have been if the rocks had been the size and weight of baseballs, as many birds are. A baseball hitting the windshield of a car moving at 65 mph would probably shatter the windshield. A bird striking the windshield of an airplane does the same thing.

Fortunately, the probability of encountering a bird strike can be greatly reduced by following a few precautions.

More than 60 percent of bird strikes occur below 500 feet. More than 90 percent occur below 3,000 feet. No altitude is completely safe. Certain migratory birds are frequently seen higher than 7,000 feet AGL. Bird strikes have been reported as high as 40,000 feet.

Peak months vary in different parts of the United States. Migratory periods in the spring and fall are usually cited as peak bird-strike times, but the bird strike "season" actually lasts from April to November. During this time, approximately 100 million ducks, 8 million geese, several hundred thousand sandhill cranes and swans, and a very large number of birds smaller than ducks migrate in North America.

In the spring, the most strikes occur from March to May; in the summer and fall, July, August, September, and October see a large number of bird strikes. The number of strikes in June and November are lower than the months just mentioned, but still substantial. Actually, the number of strikes doesn't really drop off significantly until December. To put it another way, when the weather is good for flying, birds *and* humans take to the sky.

There are several identified migratory routes all over the country. You can look these up in the *Airman's Information Manual*, if you like. The way I figure it, the birds don't know we've published specific routes for them. They probably wouldn't follow the routes if they did know. Birds are going to be flying all over the place. The pilot's best defense is see and avoid. I don't think anyone is going to alter a planned route of flight just because it crosses a migratory bird route. Merely keep a close watch when flying in the vicinity of migratory route.

You should make a concerted effort to avoid flying over protected bird refuges, however, not only because of the potential of bird strikes, but also because you could incur the wrath of an environmentalist, a fine from a judge, and possibly a violation from the FAA. Many sanctuaries have restrictions against low-flying aircraft; volunteer ground observers might be watching, ready to copy the N-number of an aircraft. Pay attention to the charts, and ask questions at local FBOs.

Birds are most active at dawn and at dusk because that's when they eat. You might want to place limitations on your flying at these times (Fig. 7-16). I hate to get up before dawn, but I love to fly then. The air is usually smooth, and watching a sunrise from an aircraft is an experience unlike any on the ground. Dusk flying is similarly romantic. I don't want to give up flying at these times just because a bunch of birds like to fly then, too. But it's something to consider.

Fig. 7-16. *The Breakfast Flock isn't too discriminating about where it eats. Trash dumps are favorite areas.*

Most of the time, birds will hear or see the airplane coming before you see them. Most birds will try to get out of the way because the airplane is obviously much bigger and noisier than they are. Most birds dive to get out of the way, so you can improve the chances of a near-miss by climbing. I've tried several times to get close enough to birds to take a photo. They always dive out of the way before I manage to click off a shot. Seagulls, it has been reported, sometimes try to outrun airplanes and often get struck from the rear. If you climb, you should be able to avoid a collision.

A few ornithologists have a theory that is unproved, but seems to make sense. Birds evolved over a very long time, and their natural enemies move at speeds no faster than 50 knots; therefore, birds survived by learning to take evasive action from anything moving that fast, mainly other predator birds. Birds that were caught did not survive to reproduce, so the ability to evade predators moving at a maximum of 50 knots emerged through natural selection.

Therefore, birds can cope with an object approaching them at 50 knots or less, perhaps faster. Around 80 knots, they can't figure out what to do (Fig. 7-17). Sometimes they

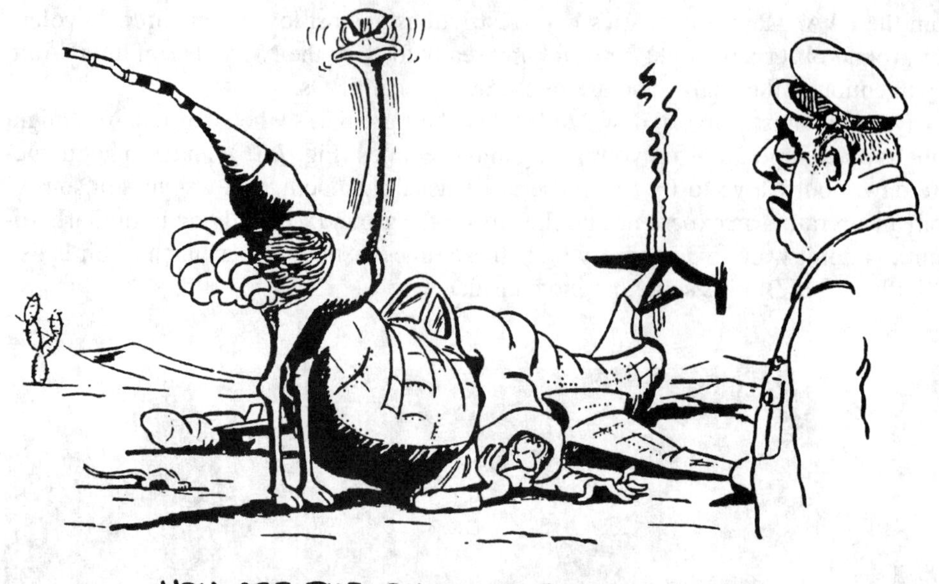

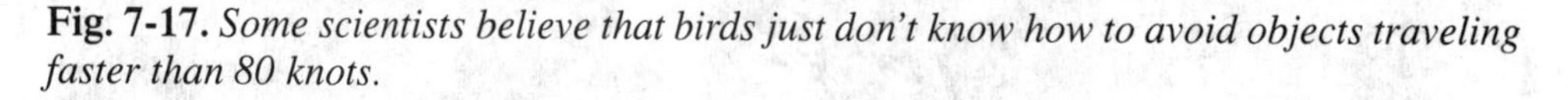

Fig. 7-17. *Some scientists believe that birds just don't know how to avoid objects traveling faster than 80 knots.*

do the right thing, and get out of the way. Sometimes they don't. They're analogous to small children who have not yet learned to judge the speed of cars when trying to cross the street; therefore, it really is up to pilots to avoid birds, not the other way around.

To increase your visibility to birds, fly with the landing lights on, particularly during departure and approach. The FAA wants you to turn on the landing light in terminal areas anyway, so you might as well do so. It couldn't hurt. Turn on the weather radar. Studies are not conclusive that either one of these works, but they might be worth a try.

If it looks like a bird strike is unavoidable, maintain steady control. Protect your face and body by attempting to duck down below the level of the windshield, which is the most common place for birds to strike. If your aircraft is equipped with windshield heat, turn it on in cold weather, even when you don't need it to clear ice or mist from the windows. (Review the aircraft operating handbook for proper operation of windshield heat.) The supplemental heat could make the difference between a bird bouncing off a pliable, warm windshield, or crashing through a cold, brittle one. That was the Air Force theory when I flew in Alaska.

After a strike, remember that aircraft control is paramount. If a bird comes through the windshield, it's going to be very messy and very smelly. Remains of the bird will probably be all over the cockpit, including you. Try not to get sick. Declare an emergency if necessary and plan to get on the ground as soon as possible. Damage to the airplane might change stalling speeds and controllability, so take the time to evaluate these at a safe altitude before committing yourself to land.

Check the engine instruments for any abnormalities. If the bird hit the engine cowling or nacelle, the cooling system might be damaged or oil lines broken. Prepare for an engine failure. If the prop is hit, be wary of unusual vibrations. Suspect possible inaccuracies from the airspeed indicator in case the pitot is clogged.

When you're on the ground, check for damage in all areas that might have been hit. Fill out an FAA Form 5200-7, "Bird Strike/Incident Report," and file it with the nearest FAA office or flight service station.

Hangar story

This isn't really a hangar story; it's a poem. And although it isn't a true story, it has a message about birds and airplanes and flying at night when strange sounds seem to magnify one's paranoia. "Raven One-One-Four" won an honorable mention in an Air Force safety writing contest. It's presented here with my sincerest apologies to Edgar Allan Poe.

By the way, a PJ is an Air Force pararescue specialist, similar to a paramedic carried on civilian air-ambulance helicopters. Years ago they were called "parajumpers"; hence, the abbreviation PJ stuck.

Raven One-One-Four

Once upon a midnight dreary, while I pondered, weak and weary,
O'er my somewhat curious clearance and forgotten safety lore,
While I nodded, nearly napping, suddenly there came a tapping,
As of someone gently rapping, rapping out a code of Morse,
"'Tis some FSS," I muttered, "tapping out a code of Morse,
Only this and nothing more."

And the sullen, sad, uncertain rumbling of each engine turbine,
Thrilled me—filled me with fantastic terrors never felt before;
So that now, to still the beating of my heart, I stood repeating,
"'Tis some FSS entreating my location to implore,
Some lone FSS entreating my location to implore,
This it is and nothing more."

Presently my soul grew stronger; hesitating then no longer,
I transmitted to the station, "This is Raven One-One-Four,
Is it I that you are calling? My reception is appalling,
And so faintly were you tapping, tapping out your code of Morse,
That I scarce was sure I heard you"—thus a voice I did implore.
Silence there and nothing more.

Deep into the cockpit staring, long I sat there wondering, daring,
Doubting, fearing violations that no pilot feared before.
But the silence was unbroken, and my navcom gave no token
As to whom or what had spoken, calling "Raven One-One-Four."
So I turned my aircraft homeward, maxing engines to a roar.
Merely this and nothing more.

Soon I was on final turning, and my eyes were burning, yearning,
For the lights upon the runway, active Runway twenty-four,
Hoping soon to be unstrapping, when I heard again the tapping,
And that unknown fearful rapping was much louder than before.
Once again I heard the tapping, which was louder than before.
"'Tis the wind and nothing more."

Suddenly there came a crashing, as if Thor were cymbals clashing;
Through my windscreen then came smashing feathers, guts, and other gore.
From my headset to my leathers, I was drenched in blood and feathers,
And in shock I could but utter, "Tower, Mayday! One-One-Four!"
That was all that I could utter, as I slumped above the floor.
Only that and nothing more.

Thinking back I can remember, it was in the cool September,
That an Officer of Safety to us pilots did implore . . .
"Beware the eagle, tern, and starling, the albatross, the swan and gosling,
For they fly with wings a-beating, and your flight will be no more,
If a fowl comes crashing, smashing through that plexiglass before,
To clip your wings for evermore."

Uncontrolled, my craft soon tumbled, and into the ground it crumbled.
So my farm was fully purchased, when they got to One-One-Four.
"'Tis no use," relayed a PJ as he came to where my form lay,
"Seems a bird ruined his whole day, for there's feathers aft to fore.
Feathers blacker than the midnight, blacker than I've seen before!"
Feathers there for evermore.

During the investigation it was found to be a raven
That had smashed into my windscreen on that fateful night of yore.
And it was upon the morrow that my friends did seek to borrow
From the bar surcease from sorrow for myself and One-One-Four,
For that fowl-destructed aircraft, call sign "Raven One-One-Four,"
Which shall fly for nevermore.

Peering now through heav'nly portals, I observe you earthly mortals,
Watching Officers of Safety while to pilots they implore . . .
"Beware the condors, hawks, and falcons, breakfast flocks of gulls and martins,
For they strike as if from cannons, and you'll slip the bonds no more."
Then, I ask my fateful raven, "Will birds ever cease to score?"
Quoth the raven: "Nevermore."

8
Human hazards

"Another twenty minutes, and I had left the thunderstorm grumbling behind. The Hudson River flowed beneath my wings, and once more I could pose as a leather-clad swashbuckler. Yet subconsciously I had discovered one of the primary truths of the aviator who might expect longevity. Prudence was necessary, but as in all things it could be ruinous if overdone. Elan was beneficial as long as it did not involve stupid sacrifice upon the altar of courage. Now I knew that the most priceless attribute of all was humility: regardless of training or equipment or time, we were all little men, self-pitted against forces often beyond human measure. Later, when my logbooks were heavy with hours flown all over the world, the lesson I had learned that tumultuous afternoon was many times confirmed: those who failed to recognize their true insignificance rarely survived."

Ernest K. Gann, *A Hostage to Fortune*

IF YOU'VE GOTTEN THIS FAR IN THE BOOK, YOU KNOW THAT I'VE TAKEN A relatively liberal approach to the definition of "adverse conditions." This chapter on human hazards, which could be subtitled, "How to look out for number one," pushes the envelope even farther.

There's an old quip that says the most dangerous part of an automobile is the nut behind the wheel. Now, I'm not implying that the most dangerous part of an airplane is the nut behind the yoke or stick, but it's been proven that most aircraft accidents, something like 70–80 percent, are caused, at least in part, by human error. This is not to say that the errors are made only by pilots. Mechanics, air traffic controllers, supervisors, manufacturers, weather forecasters, passengers, and others also make mistakes that lead to aircraft accidents. But aviation statistics show that pilots cause most of the errors that can lead to an accident.

You might not like this fact and you might disagree with it, but you have to agree that whomever it is that makes the error, it's usually the pilot who gets hurt, or worse. It's akin to riding a bicycle in traffic. Drivers can inadvertently make all sorts of mistakes: cut in front of you, open a side door into traffic without looking, run you off the road, not stop for a stop sign, and the like. The cyclist always ends up in the hospital.

So, it just makes good sense for pilots to always be on their guard. If ATC inadvertently gives you an altitude that's too low, a controller will probably feel terrible if the clearance leads to an accident. The FAA might even suspend the controller for a few months. If a lineservice attendant forgets to screw on the fuel tank cap, that attendant will probably feel really bad when the airplane runs out of fuel halfway to the destination. That attendant might get fired from the job.

An old aviation adage says the pilot is always the first to arrive at the scene of the accident. It doesn't matter who makes the error; it's up to the pilot to make sure the error doesn't result in an accident because usually the pilot's life is on the line.

There are three ways to mitigate the human hazard. The first is to take care of yourself, the second is to be suspicious of yourself, and the third is to be suspicious of others. Let's look at them one at a time.

TAKE CARE OF YOURSELF

Except for the most altruistic of us, we all tend to look out for number one. The problem is, even though we usually know what's good for us, we often don't do it. Regardless of the many reasons why we don't take good care of ourselves, there is much we can do to ensure that our bodies and minds are in the best possible condition whenever we fly.

Eat right

The old rule about eating something from the four basic food groups has changed. Now the food pyramid shows we should be eating more fruits and vegetables and fewer fatty foods. Most nutritionists recommend we eat five to nine servings of fruits or vegetables each day, and a lot less meat, sugar, and salt than most Americans have grown accustomed to.

For a few years cholesterol was public enemy number one. Now nutritionists tell us that there's both good and bad cholesterol, and a body does need a certain amount of cholesterol for normal functioning.

The good cholesterol is *high-density lipoprotein*, or HDL, which prevents coronary plaque buildup. The bad cholesterol is *low-density lipoprotein*, or LDL, which promotes plaque buildup in your arteries.

A healthy cholesterol level is under 200: the lower, the better. You should have your blood cholesterol measured at least once every 5 years, more often if you have a high cholesterol level and are trying to bring it down.

The two ways to bring down blood cholesterol are diet and drug therapies. The National Cholesterol Education Program suggests you refine your diet so that less than 30 percent of your calories come from fat, and less than 10 percent of your calories come from saturated fat.

In general, although people have different body types, men should maintain about 20 percent body fat and women about 25 percent body fat. To maintain these levels, you should eat a diet that provides you with no more than 20–25 percent of the caloric intake in the form of fat. If you want to lose weight, you need to reduce your fat intake to below these levels. Certain diets suggest fat intake be reduced to 10 percent. If you need to gain weight, you need to increase your fat intake.

Basically, there are three types of calories in food. Lean-promoting calories support the conversion of food into muscle, bone, internal organs, and essential body fluids. Fat-promoting calories support the storage of fat in the body. Empty calories, such as those found in sugar and alcohol, do neither; however, because they don't add lean-promoting calories, they should be minimized.

Much of the present American diet consists of prepared foods, and prepared foods are often high in fat content (Fig. 8-1). To be realistic, few of us can stand to subsist on fruits and vegetables, so it helps to have a way to judge how much fat intake is healthy. Fortunately, the government has mandated that processed and prepared foods carry labels listing ingredients and quantities. Using this information, it's not difficult to determine what percentage of the caloric intake of the product is supplied by fat.

Here's an example. The nutritional label on a bag of potato chips tells us that it has 120 calories and 6 grams of fat per one-ounce serving. No, you don't have to go out and buy a kitchen scale and start weighing everything you eat. All you need to know is that *each gram of fat equals 10 calories*; therefore, multiply the number of grams of fat per serving, 6 in this example, by 10, and you get 60 fat calories per serving. Dividing 60 fat calories per serving by 120 calories per serving and multiplying by 100 percent gives 50 percent.

1 gram of fat $=$ 10 calories
6 grams $\times$ 10 calories/gram $=$ 60 calories
60 calories $\div$ 120 calories $\times$ 100% $=$ 50%

This means that fat provides one-half of the calories in a serving of potato chips. (The size of the serving doesn't matter because we are figuring out the percentage.) That much fat is not a good thing. Our goal should be to eat things that have no more than 20 percent fat.

Nutrition Facts

Serving Size 1 cup (30g)
Servings Per Container About 10

Amount Per Serving	Multi Grain Cheerios	with ½ cup Skim milk
Calories	110	150
Calories from Fat	10	15
	% Daily Value**	
Total Fat 1g*	**2%**	**2%**
Saturated Fat 0g	**0%**	**0%**
Cholesterol 0mg	**0%**	**1%**
Sodium 240mg	**10%**	**13%**
Potassium 100mg	**3%**	**9%**
Total Carbohydrate 24g	**8%**	**10%**
Dietary Fiber 3g	**10%**	**10%**
Sugars 6g		
Other Carbohydrate 15g		
Protein 3g		
Vitamin A	25%	30%
Vitamin C	25%	25%
Calcium	4%	20%
Iron	45%	45%
Vitamin D	10%	25%
Thiamin	25%	30%
Riboflavin	25%	35%
Niacin	25%	25%
Vitamin B_6	25%	25%
Folic Acid	25%	25%
Phosphorus	10%	20%
Magnesium	6%	10%
Zinc	4%	6%
Copper	4%	4%

*Amount in cereal. A serving of cereal plus milk provides 1.5g fat, <5mg cholesterol, 310mg sodium, 300mg potassium, 30g carbohydrate (12g sugar) and 7g protein.

**Percent Daily Values are based on a 2,000 calorie diet. Your daily values may be higher or lower depending on your calorie needs:

	Calories:	2,000	2,500
Total Fat	Less than	65g	80g
Sat Fat	Less than	20g	25g
Cholesterol	Less than	300mg	300mg
Sodium	Less than	2,400mg	2,400mg
Potassium		3,500mg	3,500mg
Total Carbohydrate		300g	375g
Dietary Fiber		25g	30g

INGREDIENTS: WHOLE GRAIN CORN (INCLUDES THE CORN BRAN), WHOLE GRAIN OATS (INCLUDES THE OAT BRAN), SUGAR, WHOLE GRAIN BARLEY (INCLUDES THE BARLEY BRAN), WHOLE GRAIN RICE (INCLUDES THE RICE BRAN), WHOLE GRAIN WHEAT (INCLUDES THE WHEAT BRAN), WHEAT BRAN, WHEAT STARCH, BROWN SUGAR SYRUP, BROWN SUGAR, SALT, PARTIALLY HYDROGENATED COTTONSEED OIL, CALCIUM CARBONATE, TRISODIUM PHOSPHATE, CARAMEL AND ANNATTO COLOR ADDED, VITAMIN C (SODIUM ASCORBATE), IRON (A MINERAL NUTRIENT), A B VITAMIN (NIACIN), VITAMIN A (PALMITATE), VITAMIN B_6 (PYRIDOXINE HYDROCHLORIDE), VITAMIN B_2 (RIBOFLAVIN), VITAMIN B_1 (THIAMIN MONONITRATE), A B VITAMIN (FOLIC ACID) AND VITAMIN D. VITAMIN E (MIXED TOCOPHEROLS) ADDED TO PROTECT FRESHNESS.

Nutrition Facts

Serving Size 3/4 Cup (30g/1.1 oz.)
Servings per Container 19

Amount Per Serving	Cereal	Cereal with ½ Cup Vitamins A & D Skim Milk
Calories	120	160
Fat Calories	0	0
	% Daily Value **	
Total Fat 0g*	**0 %**	**0 %**
Saturated Fat 0g	**0 %**	**0 %**
Cholesterol 0mg	**0 %**	**0 %**
Sodium 200mg	**8 %**	**11 %**
Potassium 20mg	**1 %**	**6 %**
Total Carbohydrate 28g	**9 %**	**11 %**
Dietary Fiber 0g	**0 %**	**0 %**
Sugars 13g		
Other Carbohydrate 15g		
Protein 1g		
Vitamin A	15 %	20 %
Vitamin C	25 %	25 %
Calcium	0 %	15 %
Iron	10 %	10 %
Vitamin D	10 %	25 %
Thiamin	25 %	30 %
Riboflavin	25 %	35 %
Niacin	25 %	25 %
Vitamin B_6	25 %	25 %
Folate	25 %	25 %

*Amount in cereal. One half cup skim milk contributes an additional 65mg sodium, 6g total carbohydrate (6g sugars), and 4g protein.

**Percent Daily Values are based on a 2,000 calorie diet. Your daily values may be higher or lower depending on your calorie needs:

	Calories	2,000	2,500
Total Fat	Less than	65g	80g
Sat. Fat	Less than	20g	25g
Cholesterol	Less than	300mg	300mg
Sodium	Less than	2,400mg	2,400mg
Potassium		3,500mg	3,500mg
Total Carbohydrate		300g	375g
Dietary Fiber		25g	30g

Calories per gram:
Fat 9 • Carbohydrate 4 • Protein 4

Ingredients: Corn, sugar, salt, malt flavoring, corn syrup,
Vitamins and Iron: ascorbic acid (vitamin C), niacinamide, iron, pyridoxine hydrochloride (vitamin B_6), riboflavin (vitamin B_2), vitamin A palmitate (protected with BHT), thiamin hydrochloride (vitamin B_1), folic acid, and vitamin D.

Fig. 8-1. *One of the smartest things that the U.S. government ever did was mandate that processed and prepared foods carry labels giving nutritional information. The common format of these two labels from different breakfast cereals makes it easy to compare nutritional information.*

How do you figure the fat intake of food that's not labeled? Use the guidelines in Table 8-1.

Sugar and alcohol provide calories and no fat, but the calories are in essence "empty" because they don't support the conversion of food into muscle, bone, or other body elements. One of the problems with sugar is that it is often mixed with other ingredients that are high in fat content, such as milk chocolate and peanuts. Besides the obvious effects of alcohol, it also contains useless calories: a 12-ounce can of beer contains 100 calories, a glass of wine has 125 calories, and 1 ounce of distilled alcohol has 100 calories.

Table 8-1. Fat content of selected foods

Food type	Fat content
Most fruits and vegetables	less than 10%
Pretzels	15%
Grains, cereals, and pasta	less than 20%
Breads	10–20%
Most fish (excluding shellfish)	10–20%
Chicken and turkey (skin removed)	20–40%
Beef and lamb	50–80%
Whole milk, sour cream, cheese	60–90%
Eggs (all the fat is in the yolk)	70%
Pork	80%
Cookies, cakes, muffins	80–90%
Butter, margarine, shortening	80–90%
Olives, avocados, coconuts	80–90%

Using the above information, you should be able to regulate your diet without much difficulty. If you feel you need to lose weight, it's best to get a doctor's advice. For most people, losing four to six pounds a month is the optimum, both for taking it off and keeping it off.

Take a daily vitamin supplement

Much research has been conducted and is being conducted on the role of vitamins, the value of vitamin supplements, and the determination of the quantity of vitamins that we need. It's been known for a long time that the lack of certain vitamins causes diseases—British sailors were called "Limies" because they ate vitamin-C-rich limes to avoid scurvy—but there also seems to be a major shift in scientific opinion regarding the role of vitamins in disease prevention.

Because this isn't a book about nutrition, I won't delve into the subjects of antioxidants, free radicals, and other such things. If these interest you, I suggest that you research nutrition at the library or see what is available at a bookstore. In lieu of additional reading, the following advice, which we've heard time and time again, is worth repeating. Most professional nutritionists agree with this food for thought.

Unless you are specifically advised against it by a doctor, take a multivitamin and mineral supplement each day. Such a supplement should contain 100 percent of the U.S.-recommended dietary allowances (RDAs) for vitamins and minerals. This will support the nutrients you get through eating and minimize the ill effects on the days you don't get what you need.

Sleep right

Okay, you've probably heard this a thousand times: You need your sleep. Tired people make mistakes; this includes pilots.The problem is that pilots are often forced into flying when they're tired, and the FARs give fairly wide latitude when it comes to duty time, even for Part 135 and Part 121 pilots. For private pilots and others, such as corporate pilots, operating under Part 91, there are no restrictions or limits. None. Nada. Zip.

Does this mean a private, noncommercial pilot may legally fly 24 hours a day, seven days a week, 365 days a year? Yes. Does it mean that it can be done safely? No. This is one area of the FARs that allows us considerable free rein: absolute free rein, in fact.

I suspect a lot more pilots fall asleep in the cockpit than is readily admitted. I know I have fallen asleep several times, although it's always been with another pilot on board. In some flight operations, pilots have come to the conclusion that one pilot napping while the other keeps watch, and then switching after 15 minutes, is safer than two half-awake pilots. We used this procedure during early-morning flights over the North Sea when our departure time was 4 a.m., and we knew we had some 7–8 hours of flying ahead of us.

Many people have trouble sleeping in an aircraft, but I've never known a pilot who couldn't be lulled to sleep during a long cross-country flight, no matter how noisy the machine. Of course, most pilots develop a keen consciousness of an aircraft's expected noises, and any sound out of the ordinary will usually waken them in a split second.

Your goal should be not to sleep during the flight, but sleep before it. Unfortunately, that's often easier said than done, especially when we're sleeping away from home or off our normal schedule.

The following general principles about sleep might be helpful:

- A good night's rest does not equal 8 hours for everyone all the time. Perhaps 6 is enough for some people; others might need 10, or 8½, or 7, or whatever. By middle age, most people are down to about 7 hours a night, but this is by no means a standard. Some people simply need more rest; others need less. From your own experience, you should be able to determine how much is normal for you.
- A proper diet and moderate exercise during the day are conducive to better sleep.
- Most people sleep better in a dark and quiet room than in a bright and noisy room. Try to control your sleeping environment to achieve these conditions.
- Avoid eating before going to bed. The digestive process can interfere with relaxation; however, some liquids, such as warm milk and chamomile tea, help some people fall asleep.
- Stick as closely as possible to your standard going-to-bed routine. We all develop certain rituals when we go to bed. When the ritual is interrupted or altered, getting to sleep often becomes more difficult.

- Avoid sleeping pills, especially the over-the-counter kind. Drugs have different effects on different people, but by inducing sleep artificially, unwanted side effects might occur later. Always consult a doctor first, preferably an FAA-authorized medical examiner, and test the sleep-inducing pill on a night when you don't plan to fly the next day.
- Avoid using alcohol to induce sleep. Although some people swear by a nightcap, alcohol-induced sleep tends to be very low quality. Alcohol might cause you to wake up earlier than normal in the sleep cycle.
- Try to recover any sleep deficit as soon as possible. Fortunately, most people don't need to sleep an additional hour for every hour that is lost, but the longer that you wait to make up the lost rest, the more fatigued you will become each succeeding day.

Exercise regularly

The value of regular exercise has been well-proven: we sleep better, we keep our ideal weight easier, we are less stressed, and we live longer.

There are plenty of exercise books and videos around, so I won't dwell on the subject. The best exercises for long-term fitness are those that work your heart and cause you to breathe harder, the so-called aerobic exercises. Any exercise that causes your heart to beat harder for an extended period of time is an aerobic exercise. This includes walking, jogging, swimming, tennis, golf, and countless other activities.

At a minimum you should strive to exercise at least three times to four times a week using some form of aerobic activity for 30–60 minutes. Research has shown that a 45-minute walk three times a week decreases the probability of heart disease in most individuals. Make exercise a standard part of your weekly routine.

Reduce stress

We all get stressed. It's part of life. Some of us cope better under stress than others, but everyone has limits. Unfortunately, an overload of stress can adversely affect our ability to perform in the cockpit. We make mistakes; we forget things; we screw up.

Psychologists tells us that some stress is good. It gets our adrenalin pumping, and we become more alert. Performance improves. They also tell us that beyond a certain point, performance drops off. Beyond some limit, performance drops to zero. We simply give up.

Depending on how much stress you encounter, your health can be affected. There's a simple way to find out if life is piling too much stress on your shoulders (Table 8-2). This method was developed by psychologists to help predict physical problems that can result from too much change and stress in a person's life.

Table 8-2. Stress test: The Holmes/Rahe Life Change Scale

Rank	Life event	Mean value
1	Death of spouse	100
2	Divorce	73
3	Marital separation	65
4	Jail term	63
5	Death of close family member	63
6	Personal injury or illness	53
7	Marriage	50
8	Fired at work	47
9	Marital reconciliation	45
10	Retirement	45
11	Changes in family member's health	44
12	Pregnancy	40
13	Sex difficulties	39
14	Gain of new family member	39
15	Business readjustment	39
16	Change in financial state	38
17	Death of close friend	37
18	Change to different line of work	36
19	Additional arguments with spouse	35
20	Mortgage over $10,000	31
21	Foreclosure of mortgage or loan	30
22	Change in work responsibilities	29
23	Child leaving home	29
24	Trouble with in-laws	29
25	Outstanding personal achievement	28
26	Spouse begins or stops work	26
27	Begin or end school	26
28	Change in living conditions	25
29	Revision of personal habits	24
30	Trouble with boss	23
31	Change in work hours, conditions	20
32	Change in residence	20
33	Change in schools	20
34	Change in recreation	19
35	Change in church activities	19
36	Change in social activities	18
37	Mortgage or loan under $10,000	17
38	Change in sleeping habits	16
39	Change in number of family get-togethers	15
40	Change in eating habits	15
41	Vacation	13
42	Christmas	12
43	Minor violation of the law	11
	TOTAL: ______	

The Holmes/Rahe Life Change Scale. Reprinted with permission from the Journal of Psychosomatic Research, Vol. II, 1967: The Social Readjustment Scale, T.H. Holmes & R.H. Rahe; J227, pages 213-218; Elsevier Science Ltd., Pergamon Imprint, Oxford England.

To determine your level of stress, think of what has happened to you in the past year. Circle the point values for the events that apply to you. If an event has happened more than once in the last year, increase the value appropriately. Then total the numbers that you have circled. Compare the total to Table 8-3.

Table 8-3. Total from Life Change Scale test

Below 150	Little or no problem. You probably won't have any adverse reactions to the changes in your life.
150–190:	Mild problem. There's a 37 percent chance you'll feel the impact of stress with physical symptoms.
200–209	Moderate problem. There's a 51 percent chance of experiencing a stress-related illness or accident.
300 and over:	Danger! Stress is threatening your well-being. You have an 80 percent chance of a stress-related illness or accident.

If the total alarms you, do something about it. Change something under your control until your score settles down. Postpone a move or a job change. Fly under only ideal conditions. Even consider not flying for awhile.

If your total is more than 300, and you get sick, there's a good chance you'll suffer from something serious, such as cancer, a heart attack, or manic depressive psychosis. This is serious stuff. In their book, *Heart Disease: How to Work with Your Doctor and Take Charge of Your Health*, Nancy Samuels and Dr. Mike Samuels write, "Heart attacks definitely occur more frequently among people who are under stress or who have recently experienced a number of life changes that bring on a period of fatigue, depression, frustration, disappointment, and/or loneliness." If necessary, go see a doctor now and start doing something positive to lower your stress level.

If your total is less than 300, try some of these tips to reduce your level of anxiety. (You can do these things if your total is above 300, but you should also see a doctor ASAP.):

- Work it off with exercise or physical activity.
- Talk it out with someone. It helps to share your troubles with someone you trust and respect. A spouse is about the best person you can find, but a friend or family member is also helpful. A clergy person, counselor, or psychologist are other good choices. The point is: Don't keep it all inside; talk to someone.

- Learn to accept what you can't change. Some things just can't be changed. Try to live with it for now. Remember that the future will probably bring a change to the situation; conditions change, people change, lives change.
- Don't drown your sorrows with alcohol, or mask them with drugs.
- Get enough sleep and rest. (Isn't this interesting how these things are all related?)
- Balance work and recreation. We live in a dog-eat-dog, early-bird-gets-the-worm, winners-never-quit society. No wonder most of us are stressed out at one time or another. Give yourself a break and take a break.

On the same theme, a colleague of mine commented that social observers in the fifties and sixties were predicting that there would be a "crisis" caused by excessive leisure time in the next few decades as labor-saving devices freed millions of American workers from mundane tasks and reduced workweeks to fewer and fewer hours. For some reason, this never happened, at least to anyone I know.

It seems that the only people who have time to spare in the nineties are unemployed. Those of us lucky enough to have jobs seem to be working longer and longer hours. So much more reason to "take time to smell the roses." (Admittedly, I don't do it either, but I try to.) It's important to your health to have some sort of recreational activity to help decrease stress.

Control use of medications

It's a commonly held belief that medicine cures all that ails. Whether medicine is prescribed by a doctor or is an over-the-counter medication that you have selected, as a pilot you must consider the effect it will have on your performance in the cockpit.

When your doctor prescribes a medication, she should carefully explain the possible side effects of the drug. The pharmacist should do this, too, and most pharmacies print a detailed information sheet on the medication. They do this as much for their own benefit as the patient's because they don't want to be sued by someone who has an unexpected, adverse side effect. Take the time to read the information they give you. Supplement that information by researching medication references. The *Physician's Desk Reference* is commonly found at the library with other medical information books.

When you treat yourself, you become your own doctor and pharmacist; therefore, you must inform yourself of the possible adverse reactions that you might encounter.

Be aware, too, that over-the-counter drugs often don't cure the condition; they only hide the symptoms for awhile. Although this might be enough to get you through a flight, you probably won't be at peak physical performance when you fly. Worse, the medication itself might cause side effects that could affect your flying abilities.

Table 8-4 lists common over-the-counter drugs and briefly gives some of their possible side effects (FAA's *Medical Facts for Pilots*, AM-400-92/1). As with all drugs, side effects might vary according to the person, changes in altitude, and other changing flight conditions.

Table 8-4. The most commonly-experienced side effects and interactions of over-the-counter medications

	Medications	*Side effects*	*Interactions*
	Aspirin		
Pain relief **Fever relief**	Alka-Seltzer Bayer Aspirin Bufferin	Ringing in ears, nausea stomach ulceration, hyperventilation	Increase effect of blood thinners
	Acetaminophen Tylenol	Liver toxicity (*in large doses*)	
	Ibuprofen Advil Motrin Nuprin	Upset stomach; dizziness, rash, itching	Increase effect of blood thinners
Colds **Flu**	*Antihistamines* Actifed Dristan Benadryl Drixoral Cheracol-Plus Nyquil Chlortrimeton Sinarest Contac Sinutab Dimetapp	Sedation, dizziness, rash, impairment of coordination, upset stomach, thickening of bronchial secretions, blurring of vision	Increase sedative effects of other medications
	Decongestants Afrin Nasal Spray Sine Aid Sudafed	Excessive stimulation, dizziness, difficulty, with urination, palpitations	Aggravate high blood pressure heart disease, and prostrate problems
	Cough suppressants Benylin Robitussin CF/DM Vicks Formula 44	Drowsiness, blurred vision, difficulty with urination, upset stomach	Increase sedative effects of other medications
Bowel preparations	*Laxatives* Correctol Ex-Lax	Unexpected bowel activity at altitude, rectal itching	
	Antidiarrheals Imodium A-D Pepto-Bismal	Drowsiness, depression, blurred vision (See *Aspirin*)	
Appetite suppressants	Acutrim Dexatrim	Excessive stimulation, dizziness, palpitations, headaches	Increase stimulatory effects of decongestants, interfere with high blood pressure medications
Sleeping aids	Nytol Sominex	(*Contain antihistamine* Prolonged drowsiness, blurred vision	Cause excessive drowsiness when used with alcohol
Stimulants	*Caffeine* Coffee, tea, Cola, chocolate	Excessive stimulation, tremors, palpitations, headache	Interfere with high blood pressure medications

"Medical Facts for Pilots," Publication AM-400-92/1, prepared by FAA Civil Aeromedical Institute, Oklahoma City.

The brochure concludes with the following summary advice:

- Read and follow label directions for use of medication.
- If the label warns of side effects, do not fly until twice the recommended dosing interval has passed. If the label says "take every 4–6 hours," wait at least 12 hours to fly.
- Remember, the condition you are treating might be as disqualifying as the medication.
- When in doubt, ask your physician or aviation medical examiner for advice.
- As a pilot, you are responsible for preflighting your personal condition. Be wary of any illness that requires medicine to make you feel better.
- Any illness that is serious enough to require medication is also serious enough to prevent you from flying.
- Do not fly if you have a cold. Changes in atmospheric pressure with changes in altitude could cause serious ear and sinus problems.
- Avoid mixing decongestants and caffeine.
- Beware of medications that use alcohol as a base for the ingredients.

Don't smoke

If you smoke, stop. If you don't smoke, don't start. Impaired night vision is one reason that pilots should not smoke. Recall from chapter 6 that heavy smokers can lose 40 percent of their night vision at 6,000 feet due to their blood's reduced ability to carry oxygen.

Smoking in the cockpit can also be a fire hazard. The smoke residue will eventually coat the windows and gauges with a film that will reduce visibility and make the gauges more difficult to read.

There are numerous health reasons not to smoke. Heart disease is the number one killer in the United States, and cigarette smoking is the culprit in 20 percent of heart disease deaths. Nearly half the heart attacks to women under 55 are caused by smoking. The list goes on and on. Even secondhand smoke is dangerous. Chronic exposure to someone else's cigarette smoke increases heart disease by 30 percent.

So do yourself, your significant other, and everyone else in your family a favor and stop smoking.

Alcohol and illegal drugs

The FAA says no booze within 8 hours of flight. In other countries, the rule is 12 hours from "bottle to throttle." Although 8 hours is probably adequate for most people

Fig. 8-2. *The eight-hour bottle-to-throttle rule does not take into account the residual alcohol that might still be in the body.*

on most days, 12 or 16 or even 24 might not be adequate on other days. Anyone who has consumed an excessive amount of alcohol knows it usually takes more than 8 hours to feel well again (Fig. 8-2).

(If you frequently overindulge, you might be heading toward a serious alcohol addiction. Stop reading right now, go to the phone, and call your doctor for an appointment. Keep reading after you have made the appointment.)

Pilots often forget that the regulation regarding alcohol use not only gives the 8-hour rule, but also stipulates a blood-alcohol content. FAR Part 91.17 states, in part:

"No person may act or attempt to act as a crew member of a civil aircraft: within eight hours after the consumption of any alcoholic beverage; while under the influence of alcohol; while using any drug that affects the person's faculties in any way contrary to safety; or while having 0.04 percent by weight or more alcohol in the blood."

What does a blood alcohol level of 0.04 mean? Table 8-5 shows the stages of alcohol impairment. Depending on the individual, a person with a 0.04 level would fall between sobriety and euphoria. In other words, one person could show no apparent influence with slight changes in night vision, equilibrium, and reflexes. Another person might exhibit signs that are really quite dangerous in the cockpit, such as increased self-confidence, decreased inhibitions, negative changes in judgment, attention, and fine-motor coordination, and loss of equilibrium. You might argue that with a 0.04 level, you're still in the sober stage, but your night vision will be impaired and your reflexes slowed. The FAA will legally consider you under the influence.

Table 8-5. Stages of alcohol impairment

BAC*	Stage	Symptoms
.01–.05	Sobriety	No apparent influence; slight changes observed in night vision and equilibrium; reflexes slowed.
.03–.12	Euphoria	Increased self-image/confidence; decreased inhibitions; negative changes in judgment, attention, and fine-motor coordination; mild loss of equilibrium.
.09–.25	Excitement	Emotional instability; loss of critical judgment; impairment of memory and comprehension; decreased sensitory response; increased reaction time; some muscular incoordination.
.18–30	Confusion	Disorientation; mental confusion; exaggerated emotional stress; disturbance of sensation and perception of color; decreased pain sense; impaired balance; muscular incoordination.
.27–40	Stupor	Apathy; approaching paralysis; markedly decreased response to stimuli; marked muscular incoordination; vomiting; impaired consciousness.
.35–.50	Coma	Complete unconsciounsess; depressed or abolished reflexes; subnormal temperature; incontinence; embarrassment of circulation and respiration; possible death.
.45+	Death	Respiratory paralysis.

*BAC = Blood Alcohol Level = percentage of alcohol per volume of blood

National Highway Traffic Safety Administration

By the way, don't forget about another portion of FAR 91.17:

"Except in an emergency, no pilot of a civil aircraft may allow a person who appears to be intoxicated or who demonstrates by manner or physical indications that the individual is under the influence of drugs (except a medical patient under proper care) to be carried in that aircraft."

There's no such thing as a "designated flier." If your friends are drunk, you're not supposed to take them as passengers. The fact that you are the pilot and you are stone sober is irrelevant. I suspect this rule is bent to the point of breaking quite often, but it still is a violation of the regulations. If you are the pilot in command, you will be punished, not the passengers.

Alcohol's depressive effect is the same on everyone, but the rate and degree to which an individual is affected varies according to numerous factors. Among other things, the factors include the weight, height, and metabolism of the drinker; the type of alcohol consumed; the rate it was consumed; the percentage of total alcohol in the drink; what it was consumed with; if it was consumed with food; previous drinking history; and sometimes the psychological state of the drinker. All these factors notwithstanding, researchers have determined that there is a strong correlation between how much alcohol is consumed and a person's weight (Table 8-6).

Table 8-6. Estimating blood alcohol content (BAC)

Body Weight	*Number of drinks**									
	1	**2**	**3**	**4**	**5**	**6**	**7**	**8**	**9**	**10**
100 lbs.	.038	.075	.113	.150	.188	.225	.263	.300	.338	.375
120 lbs.	.031	.063	.094	.125	.156	.188	.219	.250	.281	.313
140 lbs.	.027	.054	.080	.107	.134	.161	.188	.214	.241	.268
160 lbs.	.023	.047	.070	.094	.117	.141	.164	.188	.211	.234
180 lbs.	.021	.042	.063	.083	.104	.125	.146	.167	.188	.208
200 lbs.	.019	.038	.056	.075	.094	.113	.131	.150	.169	.188
220 lbs.	.017	.034	.051	.068	.085	.102	.119	.136	.153	.170
240 lbs.	.016	.031	.047	.063	.078	.094	.109	.125	.141	.156

Hours since first drink	**1**	**2**	**3**	**4**	**5**	**6**
subtract	.015	.030	.045	.060	.075	.090

The remainder is an estimate of the percentage of alcohol in your blood.

*One drink = 12 ounces of beer; five ounces of wine; one shot of distilled liquor

National Highway Traffic Safety Administration

For example, a 100-pound person needs only one drink (12 ounces of beer, 5 ounces of wine, or one shot of hard liquor) to raise the blood alcohol content (BAC) to 0.038, which is almost high enough to preclude flying. A 160-pound person acquires a BAC of 0.023 after one drink and 0.047 after two drinks.

Recall that the body breaks down the alcohol over time. The rate of metabolization varies, but is on the average about 0.015 percent per hour; therefore, the 100-pound person in the example will metabolize one drink in about 2½ hours. The 160-pound person will metabolize two drinks in 3 hours. The FAA's 8-hour rule still holds, even if you have completely metabolized the alcohol in your blood stream.

On the other hand, let's see what happens after drinking excessively. A 100-pound person has five drinks, bringing the BAC up to 0.188. The 160-pound person has seven drinks, bringing the BAC level to 0.164. Using the same rate of metabolization, after 8 hours, the 100-pound person has a BAC of 0.068 and the 160-pound person has a BAC of 0.044.

Neither person could legally operate an airplane after 8 hours because both would still have a BAC above 0.04. To bring their respective BAC levels down to 0.04, the 100-pound person has to wait nearly 10 hours, and the 160-pound person must wait about 8½ hours.

It's not difficult to still be legally drunk the morning after. It's also important to know that coffee and a splash of cold water on the face do not counteract the effects of alcohol. They might wake you up some, but you'll still be drunk.

Is drinking and flying a problem? Although the nature of aircraft accidents usually makes it difficult to test for alcohol, many people in the FAA and NTSB believe that 25 percent of all aircraft accidents might involve alcohol. Take a look at Table 8-5 again. Are there any symptoms listed that will *improve* your ability to fly? The only reasonable conclusion is that drinking and flying don't mix.

Beyond alcohol, which you may legally consume, if you do drugs, don't fly. If you fly, don't do drugs. It's that simple.

BE SUSPICIOUS OF YOURSELF

We're entering the realm of human factors, that dark and mysterious area filled with psycho-babble like stimulus-response, man-machine interface, ergonomic factors, and other words that make many pilots wince. Actually, human factors engineers are trying to make things easier for us. The difficulty lies in the fact that we all are different.

Recall that accident statistics have proven that 70–80 percent of incidents and accidents are caused by human factors. In recent years, the aviation industry has gone through a rapid and extensive development. The products delivered from manufacturers have reached a high technical standard, and the material has become more reliable. Automatic, computer-controlled systems have become standard equipment on many aircraft.

In many ways, flying has become easier, but along with technical progress the demands on the pilot have changed in character and in some ways have become more complex.

In the early days of aviation, technical failures were a natural part of the pilot's working conditions. A pilot had to be constantly alert to anticipate and tackle problems. If you doubt this, read the works of Ernest Gann, Nevil Shute, Antoine de Saint-Exupery, Charles Lindbergh, and other early pilots.

The pilot's tasks have become monitoring, registering, and reporting. As technical progress continues, this trend will no doubt continue, too.

It seems that it has become more and more difficult to keep the pilot "in the loop," or completely aware of what's happening and what's going to happen. Staying ahead of the aircraft is just as important as it always was, but automatic systems have made it all too easy to become complacent. One reason why human factors have an increasing influence on incident and accident statistics might be the changing pilot function.

Fortunately, acceptance of the importance of human factors is increasing. Previously, one talked only of "pilot error," and it was a question of finding faults and mistakes. The tendency was to blame pilots in order to assess guilt and hand out punishments.

That did not solve the problem. Finding fault only indicates that something has gone wrong. There is seldom one factor that leads to an accident. Numerous contributing factors are usually involved.

It's important to find the reason for the occurrence. Why did it happen? Realization of the casual factors leads to a better understanding of how to introduce preventative measures. This is the role of human factors in aviation.

A brief introduction to human factors

Most of the time, people adapt well to many of the design deficiencies in their working environments, even though their overall working efficiency might be reduced. The purpose of the applied technology of human factors, or ergonomics, is to improve the efficiency of the system while providing for the well-being of the individual. When this objective is achieved, an increase in both safety and efficiency of the man-machine interface will be realized.

The field of human factors is very broad. Some of the areas of study used by human factor specialists are physiology, psychology, anthropometry, biomechanics, chronobiology, genetics, and statistics. Some of the subjects studied with specific application to aviation are operating procedures, format of manuals, checklist design, language of information, symbology, graphs and tabulation design, controls, displays, warning systems, safety equipment, seat design, cabin facilities, temperature, noise, vibration, humidity, pressure, light, pollution, circadian and biorhythmic cycles, leadership, communications, crew coordination, personal relations, and discipline.

Instead of trying to cover everything and being forced to only give you a brief taste of each area, I've decided to concentrate mainly on the man-machine interface problems. For the sake of brevity, I've divided these human factors problems into two main groups.

The first group concerns the problems associated with cockpit layout and the cockpit's switches and instruments. Remedying ergonomic problems is relatively straightforward, although not necessarily inexpensive.

The second group concerns problems that originate more with the person than with the machine. These are the mental things, the so-called "psychological baggage," that the pilot brings into the cockpit. They are not so easily fixed because diagnosing psychological baggage might be extremely difficult.

Ergonomic problems

One of the most well-known ergonomic problems that was "discovered" during World War II concerns the control quadrants of the B-25, C-47, and C-82 (Table 8-7). Pilots who flew all three types reported that they accidently cut the throttle or mixture controls when they intended to reduce engine RPM with the propeller control. Safety officers looked into the matter and quickly realized the arrangement of the controls caused the errors. A typical man-machine interface problem had been found.

Table 8-7.
Control quadrant of three WWII aircraft

Aircraft	Left	Center	Right
B-25	Throttle	Propeller	Mixture
C-47	Propeller	Throttle	Mixture
C-82	Mixture	Throttle	Propeller

You've probably run across numerous ergonomic problems in the airplanes you've flown. Perhaps you've reached down to adjust the throttle and pulled out the mixture instead. Or you wanted to turn on the landing light and turned on the cabin lights. You probably have also transferred some ergonomic problems from airplanes to cars. (Transfer problems are likely to increase when sophisticated GPS systems become commonplace in automobiles (Fig. 8-3).)

Many of these small mistakes can be attributed to a lack of standardization among various aircraft. Even airplanes built by the same manufacturer are rarely 100-percent standardized. Fortunately, aircraft manufacturers are becoming more and more concerned with human factor problems, and the cockpits designed for newer aircraft generally are better than the old ones. Of course, the newer ones also have a lot more switches, so there are many more potential problems, too.

In theory, ergonomic problems can be eliminated by redesigning the system. It might take much time and money to modify a cockpit, but it can be done. If you often flick the wrong switch in the cockpit, chances are other pilots are making the same mistake. Let the manufacturer know there's a problem, and it might be corrected in the next design.

To be realistic, however, we're probably going to have to cope with ergonomic problems for a long time, if not forever. The best way I've found to avoid flicking the

Sony Electronics Incorporated

Fig. 8-3. *Pilots will have to be watchful as technologies trickle down from airplane cockpits to automobiles. A moving map and GPS system in a car is not going to be identical to an airplane installation; therefore, a pilot will have to consciously make an effort to operate each unit differently.*

wrong switch is to look at it and confirm that my finger is on the switch I really want to manipulate before moving it. This is sometimes difficult to do at night or in the clouds. Quickly moving your head might induce vertigo, but moving your head is the only sure way to do it.

Theoretically, it is possible to eliminate all cockpit ergonomic problems; however, it is probably impossible to design a totally foolproof (pilotproof) cockpit because the second group of man-machine interface problems originates with the human in the cockpit

Psychological baggage

Every pilot who steps into the cockpit of an aircraft carries a psychological flight bag of experience, background, and conditioned responses to outside stimuli. On the surface, we might all look like we're stamped from the same mold, but inside we are all very different. We are human. And despite the concentrated efforts of instructors to standardize our behavior in the cockpit, there will always be that element of unconscious psychological control that might cause us to act in a manner diametrically opposed to what even we know is correct.

A personal example: When I first started flying helicopters, every once in awhile, while hovering, I'd press the wrong pedal when I wanted to turn. Intuitively I knew that I should press the right pedal to turn right and the left pedal to turn left, but sometimes a seat-of-the-pants reaction would cause me to press the opposite pedal first, before I could catch myself doing it. "Why?" I asked myself.

Steering a machine with my feet was an unfamiliar action, especially after I'd driven a car for some years. But it was also vaguely familiar. "Were there other things I had steered with my feet?" I wondered. Then I remembered.

Have you ever gone sledding? If you sit on a sled, you have to steer it with your feet. To turn the sled to the left, you push your right foot forward; to turn it to the right, you push your left foot forward. Being from Pennsylvania, I did a lot of sledding when I was a kid, and in the stress of learning how to hover, every now and then, my unconscious mind would take over and tell my right leg to push the nose of the helicopter around to the left.

It was a response that I had learned years before I started flying. Psychologically, my emotional state when I sledded as a kid was probably not much different from my emotional state when I was learning how to fly. Both experiences were exciting, fun, and a little scary.

Another example: When you first learned to taxi a small plane, you probably had to concentrate to remember to turn with the rudder pedals and not the yoke. But even after hundreds, or even thousands of hours, have you ever found yourself turning the yoke in the direction you want to turn while taxiing, even though your feet are doing the steering for you?

Or, have you ever caught yourself unconsciously pressing on the toe brakes of an airplane or helicopter when your final approach is a little too high and too fast? My 1946 Taylorcraft has *heel* brakes, an arrangement that I never knew about until I flew it for the first time. I have to continually remind myself to brake with my heels instead of my toes because my feet have more than 8,000 flight hours of toe-brakes to unlearn.

These are human-factor reactions that most of us overcome with habit and experience during normal operations. But when things start to get stressful, there's no telling what your unconscious might dredge up.

Compensating for every pilot's psychological baggage and stimulus-response habit pattern is impossible. That is all the more reason to get the human-factor ambiguities, the ergonomic problems, out of the cockpit. The only way is to standardize cockpits and procedures as much as possible. Additionally, an alert pilot must be constantly on guard for man-machine interface problems.

I am convinced that subtle human-factor causes are involved in most aircraft incidents and accidents. How can they not be involved? The very fact that whenever something goes wrong the pilots are in a stressful situation is reason enough to suspect that their unconscious minds influence their responses to the stress-producing stimuli.

Ergonomics and psychological baggage might become dual factors in the same incident. Consider the following case.

Hangar story

A pitch-link on the tail rotor of a Eurocopter AS332L Super Puma helicopter broke while the aircraft was outbound on an offshore flight over the North Sea (Fig. 8-4). The pilots made it safely to a ship after determining, correctly, that something was seriously wrong with the tail rotor. No one was hurt, there was minimal damage to the aircraft, the press reports were more accurate than usual, and everyone felt the pilots had done a good job. The investigation report concentrated on pitch-link and maintenance procedures without saying anything about human factors.

Fig. 8-4. *Eurocopter AS332L Super Puma helicopter.*

But listen to this. Shortly after the vibrations started, the copilot had suggested they should cut out the two autopilot lanes one at a time to see if the problem lay there. The captain, who was at the controls, looked down at the autopilot panel, and to his surprise, saw that both lanes of the autopilot were *already* disengaged. The copilot reengaged the autopilot lanes, and they both worked normally. The point is, although the captain did not consciously remember cutting out the autopilot, he obviously must have.

There are two possible reasons why he might have done this. The first reason is pure ergonomics. The second reason is more psychological.

In the first case, he might have been trying to uncouple the altitude and heading holds while still maintaining the autopilot, but accidently hit the wrong button on the cyclic, which is an error that every Super Puma pilot has made a few times. This is a typical ergonomics problem. Having both the coupler-release and the autopilot-disengage buttons on the same control invites this kind of mix-up.

The other cause might have been rooted deep in the captain's unconscious. This particular pilot had more than 14,000 flight hours in helicopters at the time of the incident, 8,000 of these in the Sikorsky S-61. The automatic flight control system (AFCS) release button on the S-61 is located in the same position on the cyclic as the autopilot-release button on the AS332 (a sensible bit of standardization); however, because a yaw problem in an S-61 could be a hydraulic servo hardover, the first emergency action is to switch off the AFCS and the auxiliary servo. This procedure was a part of the captain's active conscious for more than 8,000 flying hours. It's not hard to speculate that his unconscious mind sent a signal directly to the ring finger on his right hand while his conscious mind was telling him there was a yaw problem with his aircraft. This might have been the reason he switched off the autopilot without even thinking about it.

You might wonder how a cockpit engineer could ever design for this kind of problem. To be fair, an engineer probably can't. Recall that every pilot enters the cockpit with a vast collection of experiences and subconscious reactions. That collection puts the whole weight of the matter on the pilot's shoulders.

Eliminating human-factor errors

Pilots can do something about the matter. First, we must recognize that we are human, and we might not act under stress the way we want to or should. We have to be suspicious of ourselves.

Second, we can determine our subconscious responses to stress. A good session in a simulator will expose our gut reactions to numerous situations. Even on normal flights, you can be alert to your subconscious signals. For example, when you shoot an unfamiliar approach to minimums review the procedure. Did you forget anything, such as timing at the outer marker or verifying gear down? Whatever. Ask yourself "Why?" Then make a conscious attempt to change your habit patterns.

Finally, if you discover something in the cockpit that could cause a human-factors-related accident, let the manufacturer know. We can all tough it out and make mental notes not to get switches A and B mixed up; however, if something strange has happened to you in the cockpit, if you've ever been confused about a warning light or indication or a switch position, then the same thing has probably happened to someone else. It will no doubt happen again to you or someone else, so take time to write the manufacturer.

Three more common human factor problems

Not all human factors problems stem from the man-machine interface. Some are rooted solely in the pilot. Overconfidence, complacency, and a gung-ho attitude fall in this group.

Overconfidence. Ah, confidence. If only those people who have too little could get some from those who have too much. Psychologists say lack of confidence is one of the great inhibitors to success in this world. Accident investigators know, without a doubt, that too much confidence easily leads to destruction. The line between the two is thin, indeed.

When I was a young pilot flying Jolly Green Giants in the Air Force, one of the HC-130 Hercules pilots in our squadron told me his personal credo for staying out of trouble: "I try not to get too creative in the cockpit." I've always remembered that.

There are ways to get the job done, and there are ways to get the job done. When a pilot is flying difficult rescue missions in the harsh environment of the North Atlantic, as we were at the time, it's easy to find ways to cut corners and bend regulations. The Air Force way was often very time-consuming, but it did get the job done, most of the time. There was essentially no need to get creative in the cockpit.

The slightest hint of urgency in the mission made a lot of Air Force pilots succumb to a "regulations-be-damned-let's-get-the-job-done" attitude. I had to grudgingly admire the Hercules pilot's mature attitude. His confidence level was in perfect balance between too much and too little.

It's been said that the new pilot's enemy is inexperience, and the experienced pilot's enemy is complacency. Overconfidence, however, can hit every pilot, at any experience level. It's just a matter of degree.

If you're good at doing touch-and-goes, steep turns, map reading, or whatever, you can become overconfident about it, even if you don't have much flight time. Because high-time pilots have the experience to be better at more things than low-time pilots, the high-time pilots can likewise become overconfident about more things.

On the other hand, low-time pilots seem more prone to become puffed up with their own, albeit limited, abilities. Good high-time pilots usually know their limits much better than low-time pilots. If they're really good, they not only know what they're good at but also what they tend to get lazy about. They know the situations that are conducive to mistakes and force themselves to continually check and double-check their actions. This distinguishes great pilots from good pilots.

A Trump Air pilot I used to fly with had a trick he liked to use to catch mistakes made by him and his copilot. Several times during a flight, when no urgent tasks were at hand, he'd survey the instruments and switches before us and ask, "What's wrong with this picture?" Then we'd both look to see if there was a switch that was in the wrong position, or radio or navaid that should be tuned to another frequency, or an instrument that was giving an unusual indication. Sometimes we'd find something minor that was amiss, but most of the time we didn't. Looking was the important thing. We didn't do anything extra, but the seasoned professional had enough experience to know that all pilots sometimes have to be reminded.

One of the most difficult things about flying is knowing your own limits and the limits of your aircraft. Strictly speaking, the only way you can really know a limit is by exceeding it, then backing off by one notch. That's not a wise way to find your limits unless you happen to be training in a simulator. Of course, the manufacturer has very kindly given us the aircraft's limits for a number of things, and you would be indisputably unwise if you exceeded these on purpose.

Some limits, however, are rather vague and left up to the judgment and interpretation of the pilot. How do you know when you've reached these limits? It's impossible to know when you reach some limits. You don't know unless something breaks, and then you know that you've reached and exceeded a limit or two. The only sensible thing to do is play it extra safe by staying well inside the flight manual limitations for your aircraft.

Your own ability is even harder to gauge than the capability of the aircraft because your ability is constantly changing. As you gain hours and experience, your ability in a particular aircraft increases. You will climb a new learning curve when you switch to another aircraft, although you won't be back to square one.

Lay off flying for a few days and your skill level deteriorates a tad. Lay off a few weeks and you'll really feel rusty. Lay off a few months and you'll be embarrassed by how much you have forgotten.

Fly several hours every day for two weeks in a row and your skill level will be way up, but so will your fatigue level. At some point, fatigue will cause your skill level to drop. I almost had to fail one of the best pilots I have ever known on a company checkride because he was so fatigued from working extra days that both his judgment and his skill level were way down. He even looked very, very tired. I passed him for two reasons. First, I knew from other checkrides he could have done much better if he were rested. Second, he was going home for an extended time as soon as we were done. Failing him would have only delayed the remedy to the problem.

Your own skill level will change during a long flight as your body tires and your attention level wanes. Physiological factors like blood-sugar level, sleep (or lack of it), and biorhythms also play a part. When you think about it, it's amazing any of us ever fly safely. The only way to stay safe is to put extra limits around your limits and cushions around your capabilities. Don't fly consistently to the edge of your capability; fly inside a more limited regime. That way, when you need another "extra-something" to pull yourself out of a hairy situation, you'll have it.

Also remember that legal minimums are just that, minimums. Just because a flight legally can be done doesn't necessarily mean it can be safely done. Stack the odds in your favor by carrying more fuel than legally required. Make your personal weather minima higher than those set by the FARs, particularly when you're going somewhere for the first time. Remember that airline and commuter pilots have to be checked out over every single route they fly. That's one reason the airlines have an overall better safety record than general aviation.

"Know thyself," is one of the inscriptions at the Delphic Oracle in Ancient Greece, attributed to the Seven Sages about 600 B.C. It's not a bad axiom for pilots at the close of the twentieth century. "I think I can make it" is not so great.

Complacency. As mentioned above, complacency is usually a problem for more experienced pilots. On the other hand, I've seen some very low-time pilots who did quite well in the complacency department. Perhaps they were trying to look more experienced by emulating the complacent demeanor of another more experienced, but just as misguided, pilot they admired. Not too smart.

I always thought I knew what complacency was until I started reading related articles in aviation magazines. Apparently, the term remains ill-defined, even though most aviation professionals assume they know what it means. Nevertheless, my favorite definition is: "A conscious or unconscious relaxation of one's usual standards in making decisions and taking action." In other words, complacency is when you're so laid back you don't care what's happening.

Everyone occasionally gets in a mood like this, but in the cockpit it has proven to be a real killer. One study found that prior to a high percentage of in-flight mishaps, one or more members of the crew was heard to be whistling, as recorded by the cockpit voice recorder. Think about when your spirit moves you to whistle. Unless you're a professional whistler and perform on stage, you probably only whistle when things are hunky-dory and you feel kind of mellow.

In the Air Force, we had an expression for people who had just received their orders to be transferred to another assignment and, consequently, lost most of their motivation to do a decent job at their present assignment. They were described as "figmo," an acronym for "forget it, got my orders." It was very hard to get figmo people to do a job well. They were just plain complacent.

So what are the factors that lead pilots to become figmo? Ironically, some of the factors and conditions that are usually associated with safe flying also lead to complacency.

Ideal weather conditions, for example, tend to make us not bother about checking the forecasts. A reliable aircraft that never breaks leads us to be slack about inspections. A familiar route and a low workload can almost hypnotize us into daydreaming about other things. Did you ever miss a highway exit while driving because you were just tootling along listening to music and not really paying attention to where you were? The same thing can happen in flight.

If you fly with other crewmembers that you trust, or an experienced nonpilot passenger who helps out with the flying duties, it's not hard to let down your guard and relax a little too much. Sometimes the worst possible combination is two experienced pilots flying together. Each pilot thinks that it is OK to take it easy because the other one is so experienced.

Symptoms of complacency include:

- The acceptance of lower standards. "It's good enough for government work."
- The lack of desire to remain proficient. "I have 3,000 hours. Why should I practice?"

- Satisfaction with the status quo. "Don't rock the boat; everything is fine the way it is."
- Boredom and inattention. "I've flown this route a thousand times, why bother turning on the loran receiver?"
- Inappropriate feeling of well-being. "ATC has cleared us for the ILS. What's that song you're whistling?"
- Overconfidence and a feeling of invulnerability. "Nothing will happen to me. If it does, I can handle it."
- Preoccupation. "The autopilot has the airplane established on this straight-in ILS approach. I've never been here before, so I'll check out this airport guide find an FBO, rental car, and hotel."

Combat complacency by first recognizing that you have it, and then resolving to purge it from your attitude. Flying is a safe activity, but it does involve certain risks. The only way to minimize those risks is to stay on top of them. Face reality, and don't let complacency overcome your good sense to know that Murphy's Law can strike at any time.

Gung-ho attitude. The military is a great one for promoting the "can-do" spirit, and for good reason (Fig. 8-5). Sometimes the only way to win a battle is on the intestinal fortitude of the troops. In aviation, however, and especially in civilian flying, it's different.

Fig. 8-5. *In the military, the gung-ho spirit is often considered vital, but such a state of mind sometimes conflicts with the attitude needed for safe flight operations.*

In aviation, "can-do" and "gung-ho" must be tempered with a good dose of reality. Just because you really want to continue a flight isn't going to make the weather get better or fuel materialize in the tanks. Countless accidents have been caused because pilots thought the visibility would certainly get better ahead or that they could fly the last 5 miles to the airport on fumes. With all respect to Richard Bach's philosophy, an aircraft is a machine operating on scientific principles, not a soulful being that will go the extra mile for you just because you love it.

There's another side to the "gung-ho/can-do" spirit. For lack of a name, I call it the "will-do" spirit. Psychologists have found, not surprisingly, that many pilots often exhibit a "serve-other-people" personality trait. Medevac pilots, corporate pilots, and search and rescue pilots are particularly susceptible to allowing outside elements affect their decision-making: the sick baby, the important meeting, the sinking boat. "Get-there-itis" is another manifestation of an inflated "can-do" or "will-do" attitude.

These "will-do" pilots must continually remind themselves that these outside factors have no relation at all to the elements that the pilot should be basing all decisions upon: weather, fuel, obstacle clearances, and the like. Weather won't miraculously improve because the baby is dying. Fuel burn won't be less than normal because it's crucial for the CEO to get to a meeting. Mountains won't part just because a boat's crew is drowning.

Private pilots can easily find themselves unduly influenced by any number of outside factors: vacation running out, credit-card charges mounting, kids getting sick in the back. Anything that takes one's mind away from the principal task at hand is an outside factor. Flying is the crucial task at hand. As important as outside factors might be or seem, they must be considered secondary to the factors that have direct influence upon the safe conduct of the flight.

Don't let "can-do" and "will-do" make you do things you know you shouldn't do.

The decision is yours

Chuck Yeager, retired U.S. Air Force general and the first person to fly faster than the speed of sound, puts it this way, "If you want to grow old as a pilot, you've got to know when to push it, and when to back off."

In the final analysis, the only person who can make you a safe pilot is you (Fig. 8-6). You have to have the desire, the knowledge, the maturity, and sometimes the courage to be a safe pilot. You have to make the decision yourself. It's completely on your shoulders.

BE SUSPICIOUS OF OTHERS

Being suspicious of other people does not mean that you should be paranoid and think everyone is out to get you. You should be constantly aware that other people can do things and make mistakes that can adversely affect you. It's probably unintentional. Perhaps it's a matter of circumstance. Whatever the case, another person's mistake might be the mistake that causes your accident.

Fig. 8-6. *Know when to push and when to back off. That's the only way to grow old as a pilot.*

Human factors issues do not apply solely to the pilot in the cockpit. Stress does not occur only to the person at the controls. Studies show that air traffic controllers are subject to more stress than most pilots. Controlling a sector of airspace is like flying an ILS approach to minimums over and over again for 8 hours straight. (Controllers do take breaks, but it's still a tough job.)

Mechanics aren't usually as stressed out as controllers and some pilots; however, they often toil long hours, sometimes rush to get a job done, frequently work in poor conditions, and might not have the necessary tools. They make mistakes, too.

If you want a good example of how numerous people outside the cockpit can affect safety of flight, read Arthur Haley's book *Airport*. It inspired all the "Airport" movies and the subsequent "Airplane" comedies. The book is by far better than any of the films. It should be on the mandatory reading list of all pilots.

Remember that every individual you deal with is subject to the same human weaknesses that you have, and today might just be someone else's "bad-hair day." For your own sake, be respectfully skeptical of how well someone else is doing a job that impacts your safety (Fig. 8-7). What does this mean in a practical, nontheoretical sense?

It means doing a good preflight inspection and test flying an aircraft after a mechanic has done maintenance. (Maybe the mechanic's spouse became sick, which caused a rush job to get home sooner.)

Fig. 8-7. *Inadvertent mistakes made by other people can affect any pilot. It is up to every pilot to make sure that the aircraft is airworthy; to remain aware of the weather, aircraft position, obstacles, and traffic; fly within the limitations of the aircraft; and not run out of fuel. (U.S. Navy N2S)*

It means double-checking the weather with another source. (Maybe the person who was supposed to update the latest forecast or warning was distracted by severe weather and didn't do it on time.)

It means being aware of the minimum safe altitude for the area you're flying over. (Maybe the controller mixed up your aircraft with another aircraft.)

It means looking out for other aircraft. (Maybe the other pilot is too involved adjusting avionics to look outside and pay attention to other traffic.)

You get the idea.

9
Emergencies

"It must never be forgotten that the captain, as the pilot in command, is always responsible for the final decisions."

Helikopter Service flight operations manual

OF ALL THE POSSIBLE HAZARDOUS CONDITIONS, AIRCRAFT EMERGENCIES are probably the most feared by passengers. This fear is probably caused by several factors: human nature; book, TV, and film depictions of aircraft; and hangar stories from pilots.

Admittedly, in-flight emergencies make exciting stories that many people like to watch and hear about. Unfortunately, all this attention on emergencies has the effect of giving many folks the idea that flying is a great deal more dangerous than it really is. (My wife claims one reason that she doesn't particularly like to fly is because pilots at a social gathering seemingly only talk about the incidents and accidents that happen to themselves or other pilots.)

Recall from chapter 8 regarding human hazards that most aircraft accidents are the result of human-factor errors; however, you must be careful with a blanket statement like this. Human error often does play a role in accidents, but in many cases the errors are precipitated by an in-flight emergency or problem. Aircraft are not fail-safe. They are complicated machines that are often subjected to very hostile environments.

Things frequently break or fail on aircraft. A pilot must respond to these conditions in a way that does not turn what might be a very minor problem into a serious or fatal accident. One way to stack the odds in your favor is to use a prescribed emergency procedure. The operating handbook of an aircraft that you fly will have specifics. Beyond specific instructions you should develop a personal emergency procedure that can be applied to any situation and any aircraft.

THE BASIC FOUR-STEP AIRCRAFT EMERGENCY PROCEDURE

This procedure has been published before, but the concept is a worthwhile element of any text regarding flying in adverse conditions.

Basic Four-Step Aircraft Emergency Procedure
Step 1. Fly the aircraft.
Step 2. See Step 1.
Step 3. Memory items.
Step 4. Checklist.

If you memorize this simple procedure and follow it carefully every time you have an in-flight emergency or problem, you will greatly enhance the possibility of a safer outcome than if you did not use it.

Step 1. Fly the aircraft

With any emergency or unusual occurrence in any type of aircraft, your main concern is to keep the machine flying safely in the air until you can put it on the ground. Pay attention to heading, altitude, and airspeed.

Step 1 is your lifesaver. It is almost unbelievable that so many well-documented accidents have been caused by pilots' failure to fly the aircraft first and take care of the emergency second. The crash of Eastern Airlines Flight 401 is a well-known example.

While on a night visual approach for Miami International Airport, the four members of the cockpit crew on a Lockheed L-1011 noticed an unsafe nosegear light on the instrument panel after the landing gear had been extended. Because they wanted to troubleshoot the problem, the crew requested permission from the tower to hold west of the airport, over the unpopulated and unlit Florida Everglades. ATC approved the request.

Unfortunately, everyone in the cockpit became preoccupied with the nosegear light. Either the captain or the first officer inadvertently disengaged the autopilot. No one noticed the slow descent toward the Florida Everglades. The airliner crashed, killing and injuring many people.

Afterward, the accident investigation board determined that the gear warning was due to a burned-out bulb, but the airliner crashed because no one was flying the aircraft.

Etch step 1 into your memory right now, and remember it whenever you fly. If something happens, it should pop out and flash a distinct message to you: "Keep an eye on altitude, airspeed, and heading while you work on this problem!"

Step 2. Remember step 1

Obviously, step 2 forces you to slow down and consider step 1. Are you really flying the aircraft, or have you gotten ahead of things in your eagerness to take care of the emergency? In reality, there are very few emergencies that require a split-second response from the pilot in order to avoid a catastrophe; however, too many moments of inattention to the attitude and position of the aircraft can easily lead to what accident reports call "controlled flight into terrain."

My first experience in how not to react to an emergency happened while I was a young lieutenant in the U.S. Air Force. I was sitting in the jump seat of an HH-3E Jolly Green Giant, a large twin-engine Sikorsky helicopter, when the right engine caught fire (Fig. 9-1). A major was flying in the right seat. A captain in the left seat was designated aircraft commander for the flight. (The aircraft commander usually flew in the right seat.)

On this occasion, the captain, who was chief training officer and an instructor pilot, was checking out the major in parachute operations. The major was not an instruc-

Fig. 9-1. *Sikorsky HH-3E Jolly Green Giant helicopter.*

tor pilot. He had just arrived to the unit and was soon to be the unit operations officer. He would then be the captain's direct boss and, therefore, the one who would write his officer evaluation report. This wasn't a normal pilot and copilot relationship.

The fire warning light flashed on, the flight mechanic in the cabin said, "There's fire coming out of the right engine," and chaos broke loose in the cockpit. The captain saw the problem first, yelled "engine fire," and grabbed the cyclic and collective. The major didn't want to release the controls and started spouting out portions of the emergency checklist.

Each reached up to shut off the engine with the fire (the engine controls are on the overhead console in the HH-3E), but neither did. Hands flew all over the place. They both tried to adjust the collective to maintain rotor RPM, but when one pilot lowered the collective, the other pilot raised it. The engine indications fluctuated so much that from my position in the jump seat I couldn't tell which engine was good and which engine was bad.

After nearly a full minute of confusion, which was a very long time for an engine to burn, they finally stopped trying to do everything alone and jointly agreed on which engine to shut down and which fire extinguisher to activate. Fortunately, they chose correctly.

Everything would have been done more efficiently and safer if one pilot had concentrated on flying the aircraft (steps 1 and 2) and the other pilot had concentrated on doing the emergency procedure (step 3).

Step 3. Memory items

This step refers to the memory items in every emergency checklist that you *must know* by rote (notice I didn't say "should know") because they require *immediate action*. The initial steps of important emergency procedures are usually memory items. The steps are often printed in **bold face type** or are surrounded by a box. The purpose is the same—to make them stand out.

Most aircraft have a few emergencies that need correction so quickly there isn't time to pull out and refer to an emergency checklist. A well-constructed emergency checklist will keep "must know" memory items to a minimum because it's counterproductive if pilots are required to memorize too many things.

In many small, single-engine airplanes, the only really crucial "must know" emergency procedures deal with an engine failure and any fire, whether in the engine compartment, cockpit, or cabin. Know these by rote, and be prepared to respond anytime.

Commit the "must know" items to your memory soon after you start flying a particular aircraft. Test yourself on them often. Flying is not 100 percent excitement all the time. During the calmer periods, imagine you have an engine fire right NOW! What are the indications you can expect? What are the memory items for this emergency?

As part of the simulation, place your hand or finger lightly on the switch, handle, or control for each checklist item. This will instill in your mind an awareness of objects and actions instead of words in the checklist. Do this with all the emergencies that have memory items until you can repeat the actions without hesitation. Test yourself often enough that you do not forget.

Not all emergency procedures have immediate action items. This means a lot of things can fail that you really don't have to get too excited about. For instance, something is wrong that might require a landing "as soon as practical," but the problem is not so bad that you have to immediately start switching things on and off while calling "Mayday."

In fact, one of the worst things you can do is to start switching things on and off indiscriminately because the odds are you'll grab the wrong thing. If you don't make matters worse, you might inadvertently delay doing the right thing.

A good rule

Never switch anything on or off without first checking that your finger is on the correct switch.

This might sound very simplistic, but if you follow it religiously, you'll save yourself a lot of unnecessary trouble. Don't ever blindly reach up or reach over to switch something on or off. What you think is the heater might be the ground inverter or the landing light. You may put your finger there, but visually check it before you move the switch.

One memorable time in a simulator, I watched a senior captain with more than 10,000 flight hours very confidently reach up and shut down the good engine after the other engine had failed on takeoff. He realized his mistake as soon as he had the throttle back, but by that time it was too late. Both engines were shut down and the aircraft was flying at such a low altitude and airspeed that there was no chance to get the good engine started again.

The other pilot who was at the controls could only hold the glide to the inevitable impact with the ground. The crash would have been fatal in a real aircraft. The captain could only look at the other pilot and say, with the utmost humility, "I'm sorry. I'm so sorry." He never would have made the mistake if he had checked his hand before retarding the throttle.

This incident illustrates the value of simulator training. The captain relearned a valuable lesson that day, becoming much more aware of what he did with his hands in the cockpit. Certain instructors even advocate immediately sitting on your hands when you first discover an emergency. The reaction will keep a pilot from doing something improperly. That's okay if you have an autopilot or a two-pilot crew, but if you're by yourself you really should keep your hands on the controls. The point: "Look before you flick."

Step 4. Checklist

The last step specifically refers to the emergency checklist. After you do the memory items, if there are any, stop what you're doing, and pull out the emergency checklist. Don't do anything else until you have the checklist.

By this point in any emergency, the imminent danger has passed. The aircraft is still flying and you're in control. The immediate action items have been taken care of. Yes, there are a few more things that must be done, but you don't have to rush. Don't try to do more than the memory items, even if you think you know exactly what to do next. The potential to make a mistake is just too high. The consequence of doing the wrong thing is just too great.

Take out the emergency checklist, which should be easy to reach in the cockpit, and make sure that you find the correct malfunction checklist. It doesn't help matters if you perform the emergency items for the wrong failure. (It has been done.)

Most checklists show or tell what indications you can expect for the various failures. Confirm that you have those indications. An important word of warning: In the heat of the moment, it's very easy to misinterpret indications and see what you want to see, or expect to see. Be very critical of what you think you're seeing and double-check that you've seen it correctly. Then, if there are any memory items, confirm that you did these properly. Go on to the rest of the checklist only after you have done these two things.

Look at each item and the action required. Find the switch or lever or handle or whatever it is you need to move, and put your finger or hand on it. Confirm visually that you have the correct switch, lever, or handle. Then do the checklist task. Think about your head movements, which might induce vertigo if you're flying IMC or at night.

It might sound like a time-consuming process, but it really doesn't take much time. If you don't have time to do it correctly the first time, how will you find the time to do it again?

EMERGENCY LANDINGS

Depending on the emergency, your aircraft checklist might advise you to "land immediately," "as soon as possible," or "as soon as practical." In the case of an engine failure, you might have no choice but to land right now. This brings up the subject of emergency landings.

There are two kinds of emergency landings: precautionary and forced. A *precautionary landing* is an option chosen by the pilot instead of continuing. It could be due to a mechanical problem or outside factor, such as weather. Landing at an airport halfway to a destination to await better weather is a precautionary landing that involves no more risk than a normal landing. Landing in a field because the engine is flinging oil all over the windshield is also a precautionary landing, but this landing involves considerably more risk than a normal landing; it also has the potential of becoming a forced landing if the engine fails before touchdown.

A *forced landing* occurs when the pilot is compelled to land. In single-engine aircraft, the most common compelling cause of a forced landing is engine stoppage. Notice I said engine "stoppage," not engine "failure." The two are different, although pilots tend to lump the two together.

A failed engine breaks, stops running on its own, or must be shut down and cannot be restarted. Internal mechanical breakage of a crankshaft or connecting rod is an engine failure that usually causes the engine to stop on its own. Total loss of oil due to a blown seal is an example of an engine failure that will not stop the engine immediately, but should cause the pilot to shut it down and not attempt a restart.

A stopped engine simply stops running. It could be caused by a mechanical breakage, in which case it would also be a failed engine. It could also be caused by improper operation by the pilot, fuel starvation, or environmental factors, such as carburetor icing.

Notwithstanding engine problems, other things can cause a forced landing in an aircraft. A cabin fire is perhaps even more frightening than an engine failure. Heavy airframe and wing ice can also force an airplane down. In a helicopter, loss of main gearbox oil, tail rotor control, and severe vibrations are all valid reasons for landing immediately.

Most emergency landings in single-engine aircraft are due to engine problems; therefore, the following procedures concern emergency landings caused by engine problems. With slight modifications, the procedures can also be applied to emergency landings due to any cause.

By the numbers

You're flying along, everything is working fine, and there are no clouds in sight—BAM!—the engine makes a loud noise and stops. Suddenly, you're flying a glider. . . . Wait a minute! Is that the way engines fail?

Not often. Conceivably, a crankshaft or connecting rod could break like that, without any warning, and after a few loud convulsive turns, the engine would die. Unfortunately, the way most of us are trained—the instructor suddenly springs a forced landing on us—it's easy to get this impression of engine failures. But that's not the way engines usually conk out. Usually, there's a warning. If we heed the warning, we can often avoid the engine stoppage.

The warning will vary according to what is causing the engine to run abnormally. Statistics aren't readily available, but I would guess that the most frequent cause of a rough running airplane engine is induction icing. Recall from chapter 5 that induction icing can occur when the temperature is as high as 90°F. When you are always conscious and alert for the possibility of induction icing, and you apply carburetor heat as soon as it is suspected, you will ensure that the engine does not stop because of this minor problem.

Another common problem of a rough-running engine is an improper mixture setting. Go to full rich first, and then lean as required. Also check the magnetos. The engine should keep running at only a slightly lower RPM than normal when you switch from one magneto to the other. If RPM drops drastically or the engine stops completely when you switch to one magneto, switch to the single working magneto, and leave it there.

A restricted fuel line, or a tank near empty, can also cause a rough running engine that will be a warning to you that the engine might shut down completely. Check the fuel selector switch, and use the fuel boost pump in accordance with the airplane's procedures.

Warnings of more serious problems are unusual indications of temperature, oil pressure, and power. Temperature is a prominent indicator of trouble because heat is the prime culprit in the failure of major engine components. Excessive oil and exhaust-gas temperatures are early warnings of internal engine trouble. Low oil pressure might indicate several things, none of them good for the engine. Likewise, lower-than-normal power (engine RPM) and a rough-running engine are warnings that something is not right. You might be able to operate the ailing engine for a hundred hours, or for only half an hour.

The point is, if your engine gives a warning that it's not as healthy as it should be, have it checked by a mechanic as soon as possible. And if the warning occurs in flight, do the safe thing. Make a precautionary landing as soon as possible. Don't embark on a four-hour flight over the mountains or ocean with an engine that's telling you it's sick.

Okay, we're back in the air. Let's say you've heeded the warnings and are heading for the nearest airport, or this is one of those rare occurrences when the engine really does fail without noticeable prior warning. What do you do?

Step 1. Fly the aircraft

It's only natural that you are going to focus on that engine problem, but you must remember to maintain control of the airplane (Fig. 9-2). Before you start troubleshooting, before you start looking for a place to land, before you make that Mayday call, you must maintain and keep control of the airplane.

Fig. 9-2. *The first step in any aircraft emergency, even a complete engine failure, is to fly the aircraft while looking for a place to land.*

Because you don't know yet if you'll be able to restart the engine, or if the best landing area is far or near, the most logical configuration and speed to assume is best glide speed. This will give you the best distance and be close to the most time aloft. Trim the elevator so that the airplane will maintain this speed with little or no input from you. Got the speed? Okay, go to Step 2.

Step 2: Remember step 1

Right! Fly the aircraft. Take a deep breath and remind yourself to constantly take a moment to make sure you're maintaining the proper airspeed and are flying coordinated.

Step 3: Memory items

You have to assume the worst. The worst is that the engine will not restart. This means that you need to look for the best landing site within your glide distance.

Altitude and glide ratio are major factors in how far you can safely glide. A light airplane with a 10-to-1 glide ratio, for example, will glide 10 feet horizontally for every 1 foot of descent. At 5,000 feet above the ground, this airplane will have 282 square miles of area upon which to land, which is quite a bit of real estate. Of course, in many areas, very little of this real estate might be suitable for landing.

(To determine how much area is available for a landing from a given altitude, use the formula for determining the area of a circle: $A = (pi)r^2$, r being the distance that the airplane will glide from the present altitude. In this example, the airplane will glide $10 \times 5{,}000$ feet, or 50,000 feet. Dividing by 5,280 feet to get statute miles yields 9.47 miles. This number squared (9.47×9.47) equals 89.68 square miles; 89.68 multiplied by pi (3.14) yields 281.6 square miles.)

A pilot does have some options available to expedite the search for a suitable landing area.

First, always pay attention to your route of flight by following along on a chart. An airport might be nearby. This option is available to every pilot, all the time.

Second, ask ATC or flight service. If you're already under radar control, the controller will probably be able to give you vectors to the nearest airport.

Third, check your loran or GPS, if available. Most area navigation receivers have a function that will give the bearing and distance to the closest airport, included as a feature for just such a eventuality.

While you're looking for an emergency landing field, find out what the wind is doing on the surface. Look for flags, wash on clotheslines, even ripples on bodies of water (Fig. 9-3). Cows, I've been told, often stand with their tails pointing into the wind.

Meanwhile, you're looking out the windows, maybe talking with ATC, maintaining a good glide. It's time to see if the engine stopped because of something you can correct or if it's out for the count.

Fig. 9-3. *Knowing the direction of the wind is essential when making an emergency landing. A large column of smoke from a power plant might be one source for determining wind direction.*

Step 4: Checklist

The procedure will vary for each airplane, but generally you should change everything that has to do with the engine to another position. Switch the fuel tank selector to another tank; turn the magneto switch to all positions; adjust the mixture to full rich and throttle to full power; turn the fuel pumps on and off. Do these things in the order specified in the airplane's checklist.

If the engine starts again, level off, and try to determine why it stopped in the first place. Your concern is that it does not stop again. Depending on the cause and your circumstances, it might be most prudent to make a precautionary landing at the nearest airport.

If the engine doesn't start, you're on your way to an emergency landing. Follow the checklist procedure for shutting down the engine in flight, and turn off all unnecessary switches. Make a Mayday call, giving your call sign, position, and the nature of your emergency.

Selecting a place to land

The ideal site is always an airport runway, and an airport might be within reach, particularly if you are high. The next best thing is a surface that's hard, flat, level, and wide. The Bonneville Salt Flats in Utah or the numerous dry lake beds in the Mojave

Fig. 9-4. *Choosing the best place to land over wilderness areas is usually not an easy decision.*

Desert in California are examples; however, over most of the world, you will have to accept something less than ideal (Fig. 9-4).

Roads and highways. Often very enticing from a few thousand feet, roads and highways start revealing their hazards as you descend. Obviously, vehicular traffic is going to be a problem, particularly on busy interstates where the typical speed is in excess of 60 mph. The dilemma: Do you land going with traffic or against it?

If you land against traffic, the oncoming drivers might see you sooner, but might not respond in time to get out of the way. A head-on collision between a car or truck and an airplane is not going to be good for anyone. On the other hand, if you land going the same direction as the vehicular traffic, your airplane probably won't be seen until just before you touch down. You will have to do all the maneuvering to avoid landing on or too close to the cars and trucks; the engine will be out of commission, and you won't have much maneuvering capability.

Fields. A wide, open, smooth, grassy, and level field aligned into the wind is probably better than a road or highway. In many parts of the country, this option is rare, so you'll have to compromise, sacrificing grassy for plowed, or level for sloping, for example. With a plowed field, land parallel to the furrows, unless that would mean a strong crosswind (Fig. 9-5).

Fig. 9-5. *When landing on a plowed field, it's best to land parallel to the furrows, unless this means you must accept a strong crosswind. These fields in Pennsylvania give a pilot several choices to make a landing into the wind.*

Golf courses are a good option for airplanes with slow gliding and landing speeds, but beware the small humps and sand traps put there to catch golf balls. What appears to be level from the air might be a very bumpy fairway (Fig. 9-6).

Wooded areas. If your choice is between a boulder-laden field and woods, your best chance is probably the woods. Try to stall out just above treetop level. Aim so that you're coming between two trees instead of directly toward one. With luck, the upper branches will cushion the impact, the wings will sheer off cleanly, and the fuselage will descend to the ground without much speed.

Landing

Any forced landing, no matter how benign the terrain, should be treated as an impending crash (Fig. 9-7). Prepare by tightening your seat belts and shoulder harnesses for all occupants. Stow or secure loose objects that might become projectiles upon impact. Open a door and wedge something in the gap to prevent the door from being jammed shut. Turn off all electrical switches after making the final Mayday call. Use jackets and pillows to cushion passengers' faces, and remove eyeglasses.

The operating handbook's emergency landing procedure for one single-engine airplane provides an example:

"When assured of reaching the landing site selected, and on final approach:

1. Airspeed—ESTABLISH NORMAL APPROACH SPEED.
2. Fuel selector valve—OFF.
3. Mixture—IDLE CUTOFF.

4. Magneto/start switch—OFF.
5. Flaps—AS REQUIRED.
6. Battery, alternator, and fuel boost switches—OFF.
7. Cabin-door latch—OPEN."

Flaps and landing gear can be used at the last moment to slow the airplane when the landing field is within reach. If the airplane has retractable landing gear, consider landing with the gear up. This will shorten the landing roll considerably, although it will cause damage to the prop and fuselage. If the landing field is very soft or bumpy, landing gear-up might avoid nosing over at a high rate of speed, causing even more damage.

Fig. 9-6. *A golf course can sometimes be an acceptable landing area for light airplanes, but watch out for sand traps, small hills, and golfers.*

After landing

Clear away from the aircraft quickly. Depending on the severity of the crash landing, a post-crash fire is likely. Attend to any injuries and start thinking about survival, if you're way out in the boonies.

If you're not way out in the boonies, try to get to a telephone as soon as possible to let flight service or ATC know that you're on the ground and what kind of assistance you require, if any (Fig. 9-8).

Fig. 9-7. *Treat every emergency, power-off landing as an impending crash and prepare by tightening your seat belt and shoulder harness, opening a door and wedging something in the gap to prevent it from being jammed shut, and declaring an emergency.*

Robert Hoffman/The Bear Valley Voice

Fig. 9-8. *After leaving the aircraft, try to contact ATC as soon as possible to inform them of your condition and what assistance you require.*

DECLARING AN EMERGENCY

Many pilots, if not most, seem reluctant to declare an emergency, even when they feel they really could use the help. This reluctance often stems from a fear of punishment by the FAA, in spite of the agency's frequent assurances that it won't punish pilots who declare a bona fide emergency. Of course, most of us take that with a grain of salt, given the stories we've heard almost anytime pilots get together.

Hangar stories

Larry Bothe, a pilot with more than 2,700 flight hours, has been in four declared emergencies. He was pilot in command for three of them, and a passenger during the other one. Of the three when he was PIC, he declared two and the other was declared for him by the tower. He has consented to relate his experiences here because they contain some lessons that might be useful to pilots who find themselves faced with the decision to declare an emergency or not. The following are his own words.

Iced up over a Navy base. My first declared emergency was weather-related. I had a bad case of "get-home-itis," and was flying in very marginal VFR conditions in snow showers. As I headed south, it warmed up a bit, and I got into freezing rain. I was down to 1,000 feet AGL and in the airport traffic area of a naval air station when I decided to call it quits. Because I was already on their frequency, it was easy to just say, "I have an icing emergency and want to land." That's all I did. The tower instructed me to give fuel remaining and souls on board. My answer was "two hours and one soul," and I was cleared to land.

After landing, I was met on the ramp by a tough chief petty officer who was obviously very unhappy about coming out in the cold rain to park a civilian airplane. I was very glad to be on the ground, but I figured I was in big trouble. Once inside the operations building, I was greeted by the much more cordial officer in charge who gave me coffee and let me use the phone to call my wife to have her come get me. During the wait, I finally summoned the courage to ask what was going to happen to me as a result of declaring an emergency and landing at a military base.

The surprise answer was "Nothing." The officer in charge said that because I had followed the correct procedures and had not caused any damage, he would not report it to anyone. He did say that had I just appeared on his runway with no prior communications, it would have been a different story. He also mentioned that he had a particular dislike for flight instructors whose students land without calling because they are lost and have no idea where they were. This apparently happened about once a week at the base and he always reported it to the FAA for appropriate action.

Engine failure imminent. The second emergency happened a few years later. A friend of mine needed a ride to Philadelphia International Airport (PHL) one rainy Saturday. I filed a flight plan from home and drove to the airport. When I got there, my plane wouldn't start.

Fortunately, one of the local instructors was trying to set up some actual IFR training with a student, and he volunteered to take us to Philly as a training exercise. Because I was already on file, we climbed into the instructor's Cessna 172, called for a clearance, changed the aircraft number, and were off.

The trip there was uneventful, but the departure from PHL for the return was a different matter. Passing through 5,000 feet in clouds, the instructor motioned for me to look at the engine gauges. I was in the backseat with no intercom. Oil-pressure and temperature needles were both in motion, going down and up. Because I figured all the oil was running out of the engine, I pointed toward the ground.

The instructor called Philadelphia Approach and declared an emergency, explaining that engine failure was imminent. Approach control immediately assigned us a lower altitude, which the instructor refused. Because we were several miles from the airport, he wanted to maintain as much gliding distance as possible.

We were vectored to the field and cleared to spiral down through the clouds. Upon reduction of power, the oil temperature stabilized at the very top of the red, and the oil pressure fell to about halfway between the bottom of the green and zero. We broke out at 1,300 feet agl and saw flashing lights all over the field. I never knew PHL had so many fire trucks. (There were 11, six on one side of the runway and five on the other.) Fortunately, we landed without incident.

An airport administrator with an airline mentality gave us the devil for disrupting operations at his airport, but all we ever heard from the FAA was a phone call a few days later to get the instructor's version of what happened.

We never found out what caused the strange engine indications. There weren't any signs of a massive oil leak, although the engine oil quantity was right at minimum. We added 2 quarts, did a full-power runup, which was okay, and flew home without problem. The next day the mechanic couldn't find anything wrong, and the problem never reoccurred.

Alternator problems. Flying to Florida at Christmas one year, I had an alternator failure over Richmond, Virginia. The weather was excellent. I was already talking to approach control while I was transiting the area, so I just told them I wanted to land at Byrd Field. I also casually mentioned that my alternator had quit.

As I turned final, the fire trucks were on the runway, two of them this time. They followed me all the way to the FBO, as if the airplane might explode or something. It was actually somewhat embarrassing.

I walked inside the FBO and called the tower to find out what all the fuss was about. I was told that whenever a pilot reports an equipment failure, tower personnel are required to declare an emergency on the pilot's behalf if the pilot has not declared one.

I ended up staying overnight while an alternator was flown in from Raleigh. The next day, Christmas Eve, when I went to see how the repairs were coming, I was met by an FAA inspector who wanted to ask me some questions.

The mechanics discovered that the failed alternator on my plane was for an automobile, not an airplane, and the inspector wanted to know how it got there. I told him it was news to me because I bought the airplane used, and the alternator was installed when I bought it. There was no logbook entry showing it had been replaced, so the inspector dropped the matter. That was all I heard about that emergency.

Engine problems again. I bought into a Cessna 182 and was flying to Annapolis, Maryland, to pick up a friend. The weather was marginal, but I was running late, so I didn't file IFR. After departure, I found myself between layers in light rain. It was dark enough that I was having trouble reading the loran, so I turned up the panel lights. As I fiddled with the lights and the loran, I suddenly became aware that something was wrong. The oil-pressure and temperature gauges were in motion, not unlike the earlier incident over PHL.

I squawked 7700 and declared an emergency on 121.5. Philadelphia Approach responded and told me that West Chester Airport was at a heading of 330° and only 4 miles. That confirmed what the emergency search feature of the loran had told me. I was already turning in that direction.

I didn't believe the engine would run long enough for a full approach, so I waited until I was about two miles from the field, pulled out the throttle, and descended into the undercast. I broke out about 1,000 feet AGL with the field straight ahead. The landing was uneventful.

I went to a pay phone to call flight service so that they could relay to Philadelphia Approach that I was safely on the ground. I had once again disrupted all the traffic around PHL; I wanted to release the airspace back to normal. I went back to the airplane to try to find out the problem.

I checked the oil; it was full. I started the engine; it ran normally and oil pressure and temperature were correct. I wasn't completely satisfied, so I hunted for a mechanic. I found one, and even knew him vaguely. He ran the engine at different power settings and observed the instruments. Everything looked normal.

We found the problem when he turned up the panel lights to see better. Varying the dimmer control of the panel lights made the engine's temperature and pressure gauges say anything you wanted them to. Satisfied that the engine was okay, I continued my trip without using the panel lights. The FAA never called me about the declared emergency.

A short time later an electrical specialist looked over the airplane. After rewiring some circuits, the panel lights worked, the radio noise went away, the transponder dimmer worked, and the engine instruments didn't give false readings.

Lessons learned. So there you have it. Four emergencies that were formally declared over a period of 18 years. None of them resulted in any action against me by the FAA. I never even had to write a single word of explanation. Maybe the FAA isn't so bad after all, or maybe there are other reasons the FAA never called me.

I believe there are two. First, I followed the correct procedures in conducting the emergencies, both in the initial declaration and in the actions immediately after the danger passed. For example, if I had not called flight service as soon as I got on the ground after the last emergency, then all approaches into PHL would have been halted until they found me and made sure that I was no longer wandering around in the clouds. Any significant delay at a large international airport would surely have warranted an investigation.

Secondly, I landed without incident in all cases. I didn't break anything. If I had done any damage to the airplane or anyone had been injured, I'm sure an investigation would have ensued. There were no damages or injuries probably because the emergencies were declared early, before any of the situations became unmanageable.

Based on my experience, I would encourage pilots to declare an emergency at any time it becomes apparent that safe flight can no longer be continued. *Do it before things really get out of hand.* As long as you do it correctly, don't have an accident, and haven't done anything blatantly wrong, [apparently] the worst thing that will happen is a phone call from the FAA for your side of the incident.

Just have the presence of mind to do the right thing when you get on the ground so that the declared emergency doesn't tie up the surrounding airspace for a long time. A thank-you call and follow-up letter to the facility that assisted you won't hurt either; it's the courteous thing to do, as well.

In the future, I might not mention an alternator failure on a CAVU day, but I certainly won't hesitate to take advantage of assistance from the ground anytime that I think it might help.

10 Postflight

"Once a man becomes committed to a life in contest with the elements he cannot ever again ignore them. Thus do old sailors and aviators continue to observe the constant changing of the sky, sniffing at the winds, eying the texture and formation of clouds, the quality of visibility, and all the other nuances which in the past have been familiar and have helped him survive."

Ernest K. Gann, *A Hostage to Fortune*

I HOPE YOU'VE ENJOYED READING *FLYING IN ADVERSE CONDITIONS* AND found some things that will help you when the chips are down. If you have any questions or comments, please feel free to write to me in care of TAB Books, a division of McGraw-Hill, Inc. I'll do my best to reply to you as quickly as I can.

And if you have any of your own hangar stories that you'd like to share, I'd like to hear them. You would be granted anonymity upon written request, but I would require your name and address for verification. (Who knows? Your story might end up in a subsequent edition of *Flying in Adverse Conditions*.)

I also encourage you to read other books about weather and flying. Although the elements remain the same as they have since the dawn of aviation, our collective knowledge of meteorology is improving all the time.

I would like to conclude with three short points.

First, "A little knowledge is a dangerous thing." Although there are ways to cope with most adverse flying conditions, sometimes the best way is to just say "no" and stay on the ground. Don't let the knowledge you glean from this book and from your own experience make you too cocky. Achieving the status of holding a pilot certificate reflects your ability to practice sound judgment. Never hesitate to wisely utilize that judgment for your sake, the sake of your passengers, and the sake of loved ones. Play it on the safe side, always.

Second, what your high school driver's education teacher told you about driving applies tenfold to flying: "Always leave yourself an out." You never know when you'll need it.

Third, when all else fails, you can always try a pilot's weather forecasting stone, which I guarantee will give you years and years of fantastic maintenance-free "now-casting." Instructions are in Table 10-1.

Fly safe.

Table 10-1.
Pilot's weather forecasting stone

Condition	Forecast
Stone is wet	Rain
Stone is dry	Not raining
Stone casts a shadow	Sunny
Stone is white on top	Snowing
Can't see the stone	Foggy
Stone is swinging	Windy
Stone jumping up and down	Earthquake
Stone gone	Tornado

Appendix A
Survival kit

PREPARING A SURVIVAL KIT WILL PAY OFF IN WAYS THAT YOU'LL NEVER realize until you need one. These suggestions are ideal for a private aircraft that is flown primarily overland.

Container

Any lightweight metal or plastic container with a lid, suitable to carry and store water.

Tools

- Hacksaw—single handle with blades for wood and metal
- Short-handle axe
- Pliers—vise-grip
- Pliers—slip joint
- Screwdriver set
- Leather work gloves

First-aid kit (one person)

- Sealable plastic container
- 2 compress bandages

- 1 triangle bandage
- 1 small roll 2-inch tape
- 6 3-×-3-inch gauze pads
- 25 aspirin
- 10 Band-Aids
- Razor blade or scissors
- Hotel-size soap bar
- Kotex—purse size
- Kleenex—purse size
- 6 safety pins
- 1 small tube of Unguentine or Foile

Shelter (one person)

- Two 9-×-12-feet heavy-gauge plastic sheets (red or yellow for use as signal panels) or one good-quality mountain tent with rain fly

Clothing

- Dress for the worst conditions expected
- Sturdy hiking shoes or boots
- Extra pair of socks
- Long trousers, wool in colder areas
- Long-sleeved shirt, wool in colder areas
- Wool sweater or down-filled vest
- Rain suit
- Mountain parka of wind-resistant, water-resistant material
- Hat or wool cap
- Gloves or mittens
- Mosquito netting to cover head

Food and energy (one person)

Five-day rations for one person. Put each item in a small plastic bag and seal. Put everything in a small metal can (cooking pot) and seal with a plastic bag and tape.

- 3 cans Sego, Nutriment, or Metrecal
- 30 sugar cubes
- 10 pilot bread or 25 crackers
- 10 packets of salt
- 3 tea bags
- 5 sticks of gum
- 10 bullion cubes
- 20 protein wafers
- 1 collapsible water container (optional)
- 1 small backpacking stove with sealed fuel containers (optional)

Life Support Kit (one person)

- Waterproof matches
- Candles or fire-starter
- Signal mirror
- Good quality compass
- Pocket knife—Boy Scout or Swiss Army type
- Insect repellent
- Mosquito netting to cover body
- ⅛-inch nylon rope (50 feet)
- Whistle
- Smoke flares or red day/night flares
- Nylon nonrigid daypack
- Sleeping bag with waterproof ground sheet (optional, but very nice to have if you must spend the night outside)

Table A-1. Life expectancies when requirements for life are not met

Required for life	You can live without for about
Air	3 minutes
Shelter	6 hours in severe weather
Water	3 to 6 days
Food	3 weeks
Will to live	?

Appendix B
Weather-related accidents and incidents

IN 1993, A TEAM FROM THE FAA'S OFFICE OF AVIATION SAFETY LOOKED AT aviation data from a variety of sources including NTSB accident data, FAA pilot deviation data, FAA near mid-air collision data, air traffic flight assist data, and NASA Aviation Safety Reporting System data. The examination focused on weather-related issues.

Over the 1988–1992 time period, the annual number of weather-related accidents declined; however, the annual number of weather-related accidents remained constant as a percentage of total accidents. The annual number of weather-related incidents also declined. NTSB data indicate that approximately 55 percent of all weather-related accidents show no record of a weather briefing.

Between 1988 and 1992, there were 12,391 aircraft accidents. Weather was a contributing or causal factor in 2,683 (22 percent) of these accidents. General aviation aircraft were involved in 2,347 of these weather-related accidents. The weather factor that was cited varies, depending on the category of operator.

Of the 2,347 weather-related accidents involving general aviation aircraft, 1,649 (70 percent) fall into the "personal" category. The major weather factors in this category were:

- Winds (42 percent)
- Visibility/ceiling (25 percent)

- Density altitude (8 percent)
- Turbulence (8 percent)
- Icing (7 percent)

Of the 2,347 weather-related accidents involving general aviation aircraft, 296 (13 percent) fall into the "instructional" category. The major weather factors in this category were:

- Winds (59 percent)
- Icing (13 percent)
- Density altitude (8 percent)
- Turbulence (7 percent)
- Visibility/ceiling (7 percent)

Of the 2,347 weather-related accidents involving general aviation aircraft, 223 (10 percent) fall into the "business" category. The major weather factors in this category were:

- Visibility/ceiling (35 percent)
- Winds (26 percent)
- Precipitation (12 percent)
- Icing (10 percent)
- Turbulence (9 percent)

The most prevalent cause factors in fatal accidents were:

- Visibility/ceiling cited 672 times
- Icing cited 45 times
- Thunderstorms cited 41 times

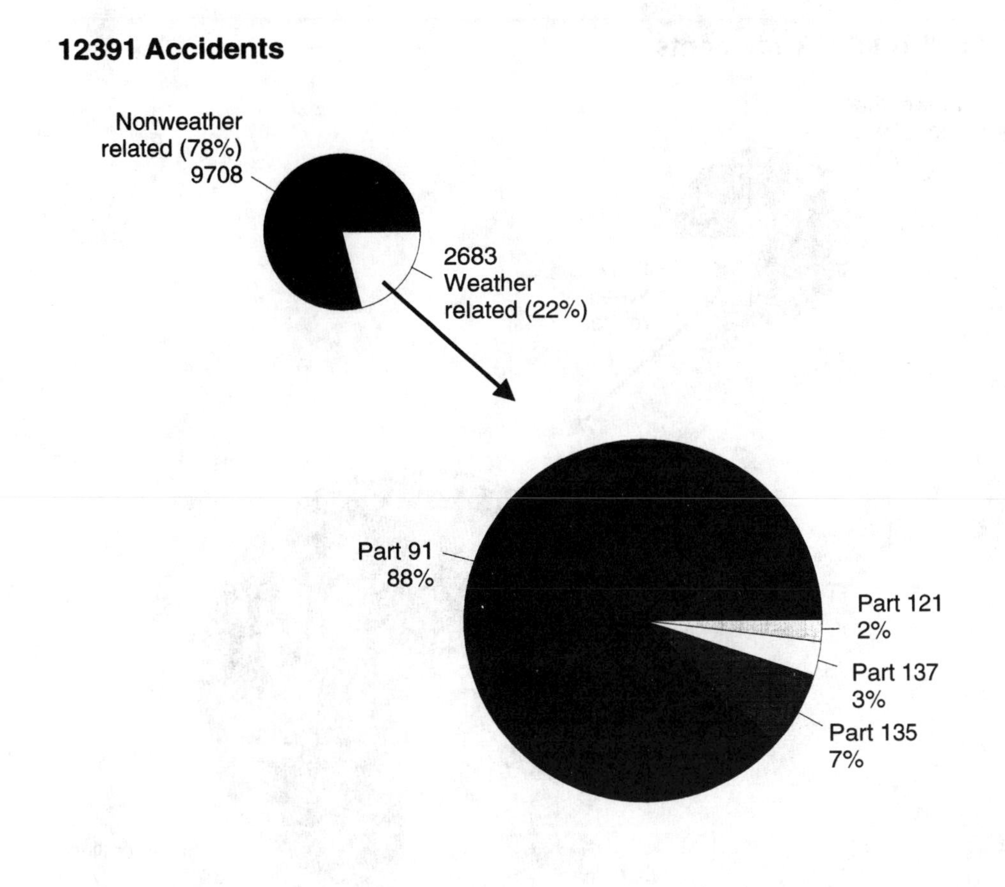

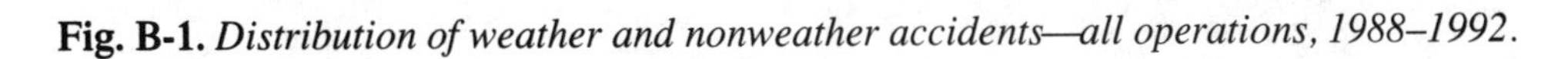

Fig. B-1. *Distribution of weather and nonweather accidents—all operations, 1988–1992.*

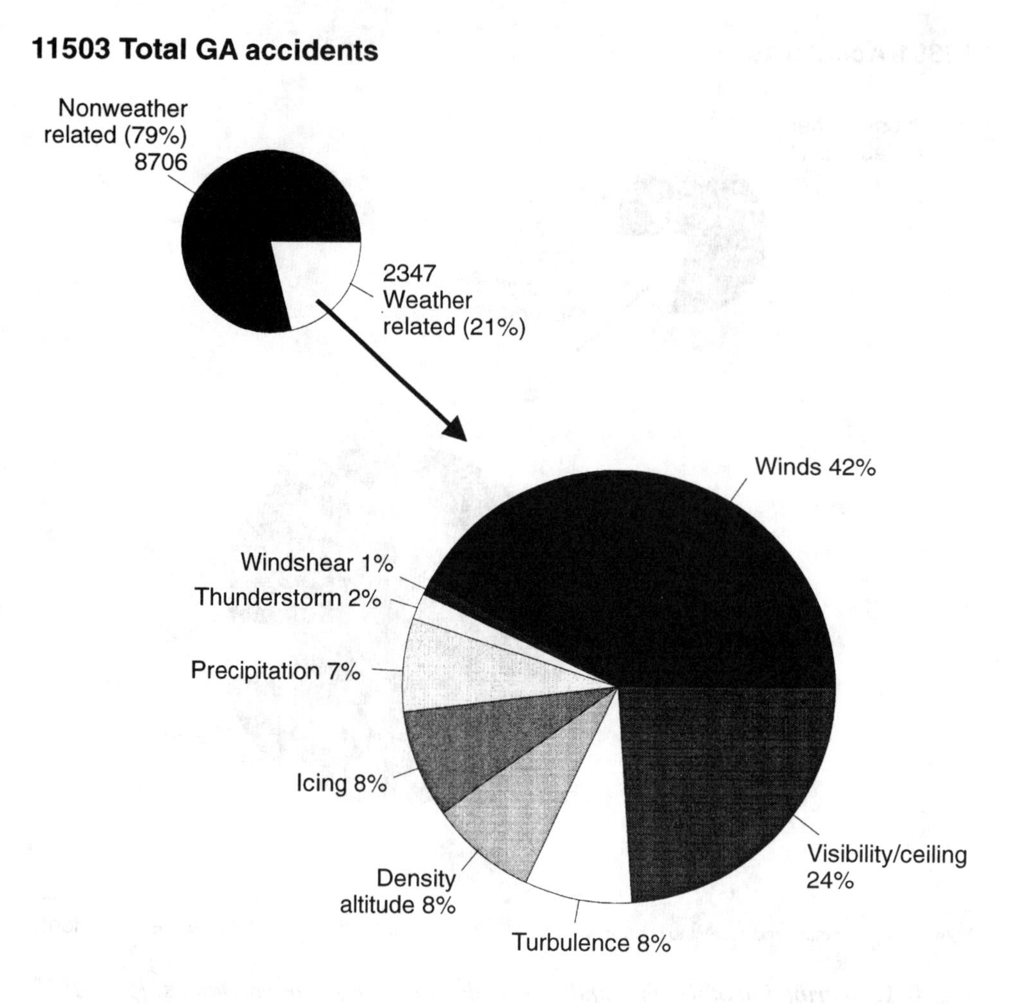

Fig. B-2. *Distribution of weather and nonweather accidents—general aviation only, 1988–1992.*

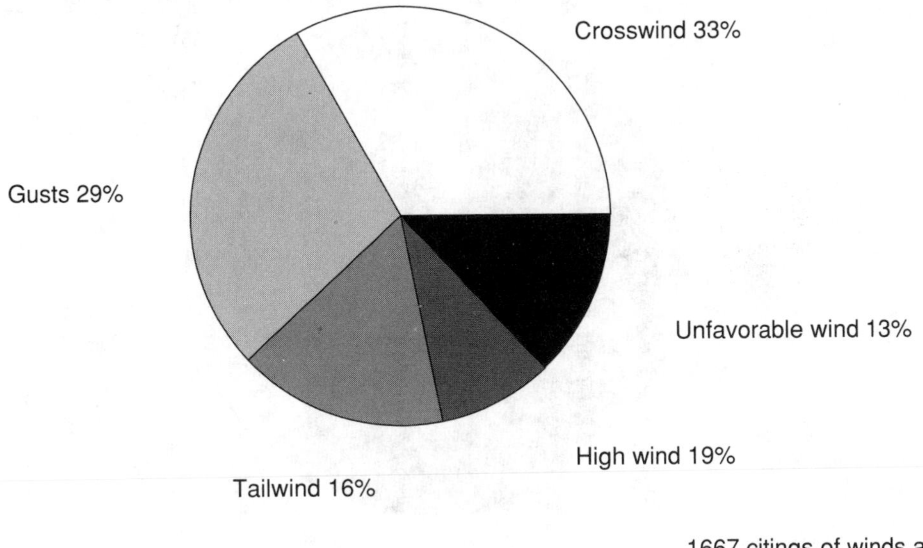

Fig. B-3. *Distribution of wind-related weather accidents—all operations, 1988–1992.*

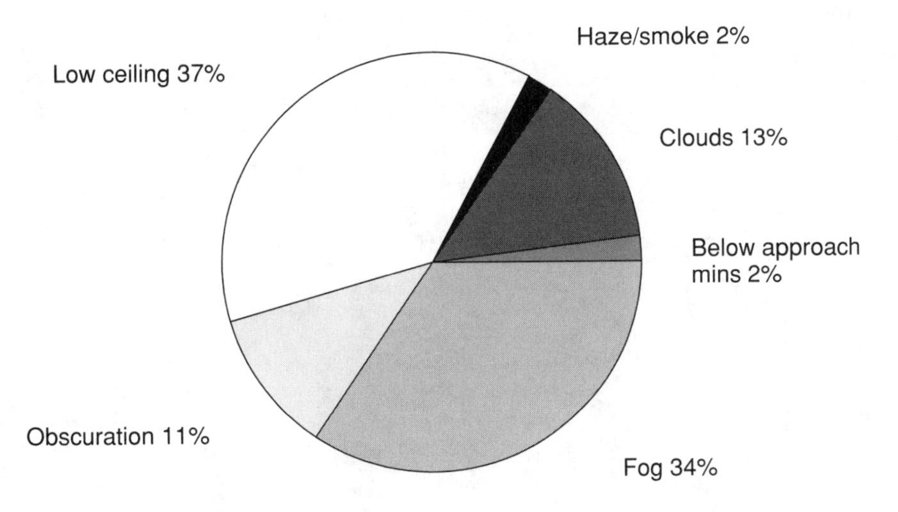

Fig. B-4. *Distribution of visibility/ceiling-related weather accidents—all operations, 1988–1992.*

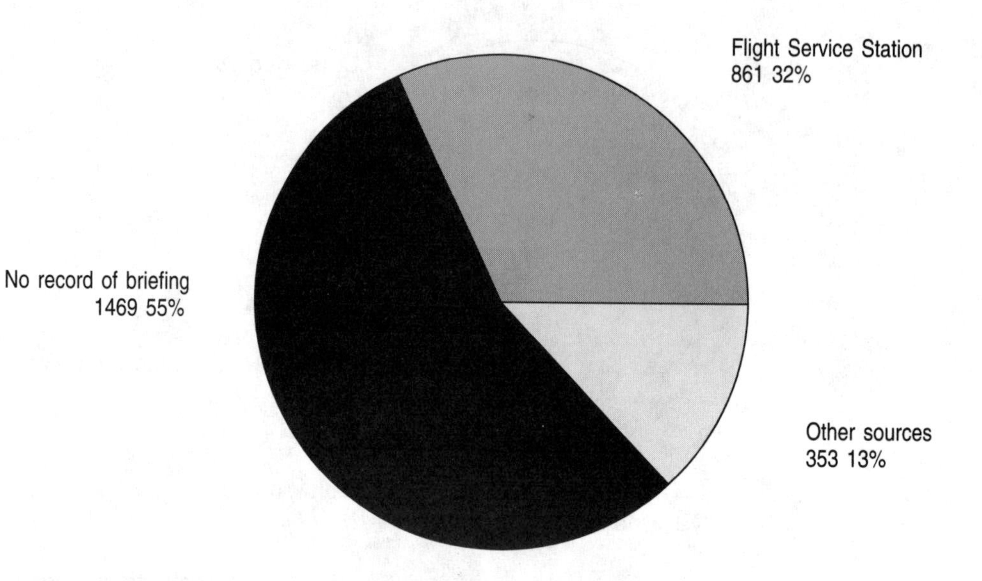

Fig. B-5. *Weather briefing sources of flights involved in weather-related accidents—all operations, 1988–1992.*

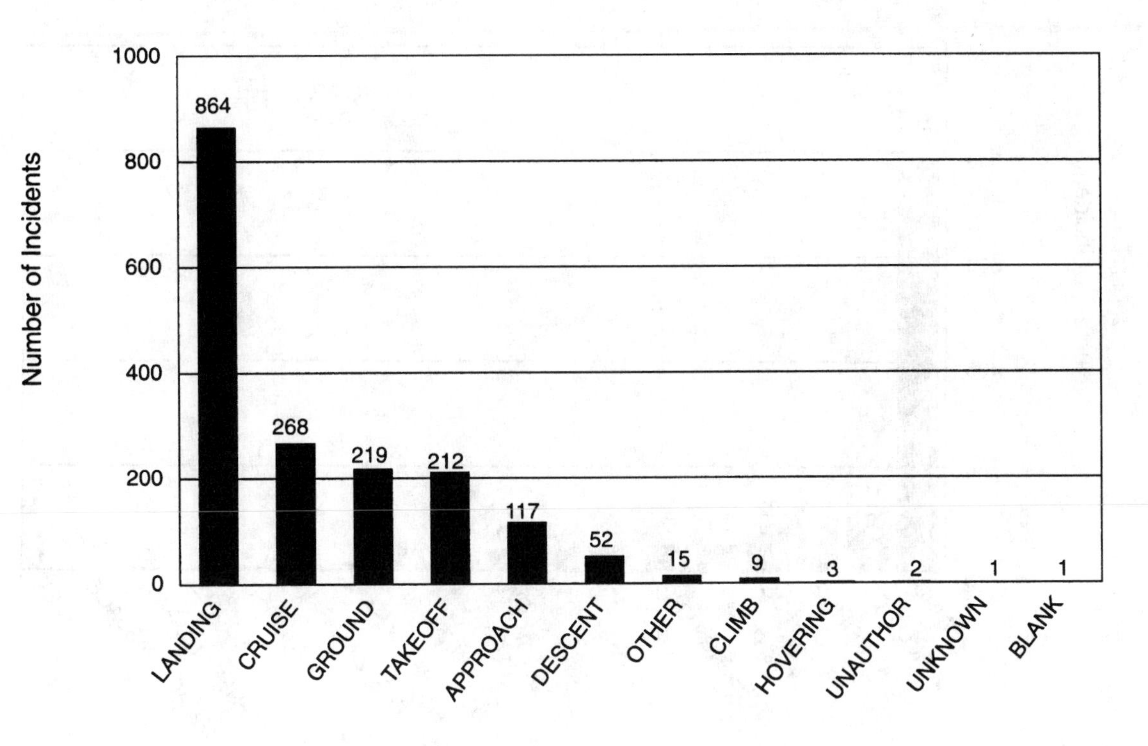

Source: Accident/Incident Data System
Note: Data are preliminary and subject to change

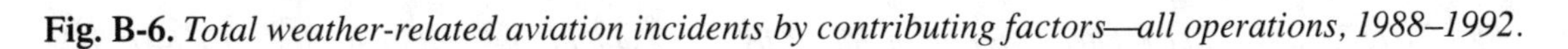

Fig. B-6. *Total weather-related aviation incidents by contributing factors—all operations, 1988–1992.*

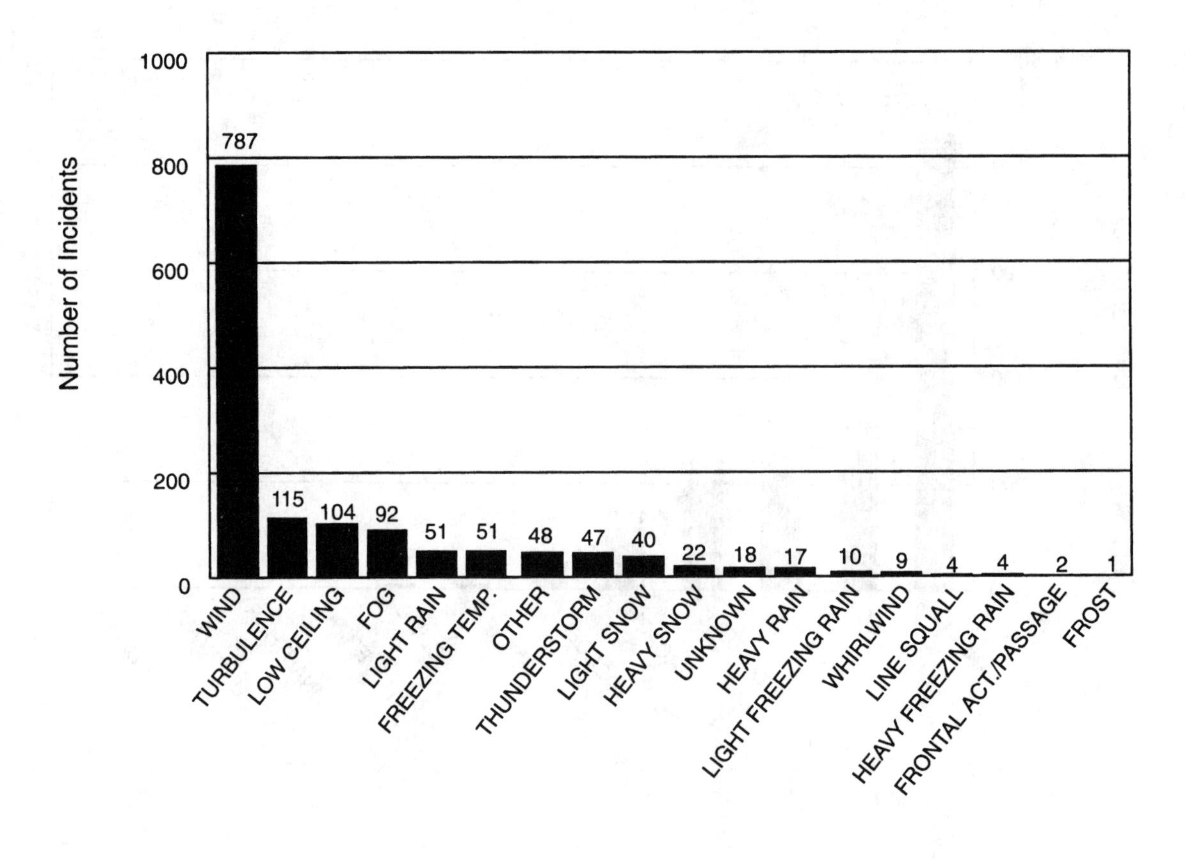

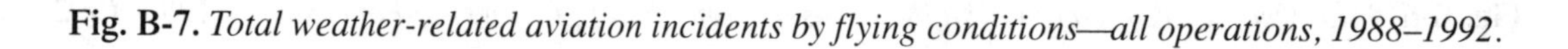

Fig. B-7. *Total weather-related aviation incidents by flying conditions—all operations, 1988–1992.*

Glossary

This glossary contains important terms used in text, plus other weather-related terms that are not found in the book but are often used in weather forecasts and other materials concerning meteorology.

above ground level (agl) The height of an object above terrain, rather than above sea level.

absolute humidity In a system of moist air, the ratio of the mass of water vapor to the total volume of the system; usually expressed as grams per cubic meter.

accretion Growth of a cloud or precipitation particle by the collision and union of a frozen particle with a supercooled water droplet.

active front A front that produces much cloudiness and precipitation.

adiabatic process The process by which fixed relationships are maintained during changes in temperature, volume, and pressure in a body of air without heat being added or removed from the body.

advection fog (or **sea fog**) Fog formed by warm, humid air flowing over cool ground or water.

Aircraft Owners and Pilots Association (AOPA) An association that represents and supports the needs of general aviation pilots.

air defense identification zone (ADIZ) The area of airspace overland or water, extending upward from the surface, within which ready identification, location, and control of aircraft are required for the interest of national security.

Airman's Information Manual (AIM) A primary FAA publication whose purpose is to instruct airman about operating in the National Airspace System of the United States.

air mass A large body of air with relatively uniform characteristics, such as temperature and humidity.

airmen's meteorological information (AIRMET or airmet) An advisory of potentially hazardous conditions, mainly of concern to small aircraft and are issued as necessary except when already part of an area forecast.

Airport/Facility Directory (A/FD) A publication designed primarily as a pilot's reference manual containing all airports, seaplane bases, and heliports open to the public including communications data, navigational facilities, and certain special notices and procedures.

airport surveillance radar (ASR) Approach control radar used to detect and display an aircraft's position in the terminal area.

air route surveillance radar (ARSR) An air route traffic control center (ARTCC) radar that is used primarily to detect and display an aircraft's position while en route between terminal areas.

altimeter A special type of aneroid barometer used in airplanes to measure altitude.
altimeter setting The value of atmospheric pressure to which the scale of a pressure altimeter is set.
altocumulus White or gray layers or patches of cloud, often with a waved appearance. The cloud elements appear as rounded masses or rolls, composed mostly of liquid water droplets that might be supercooled and might contain ice crystals at subfreezing temperatures.
anabatic wind An upslope wind due to local surface heating.
anemometer An instrument used to measure the force or speed of wind.
aneroid barometer A barometer that measures atmospheric pressure using one or a series of aneroid capsules.
aneroid capsule A thin metal disc, partially evacuated of air, that is used to measure atmospheric pressure by measuring its expansion and contraction.
anticyclone An area of high atmospheric pressure that has a close circulation that is anticyclonic (clockwise in the Northern Hemisphere and counterclockwise in the Southern Hemisphere).
anvil cloud The popular name of a heavy cumulonimbus cloud having an anvil-like formation of cirrus clouds in its upper portions.
AOPA Aviation USA A comprehensive publication of AOPA that includes landing facilities (airport, seaplane base, and heliport information), aircraft and avionics directories, and a "pilot's source book" of aeronautical information, regulations, procedures, and reference material.
atmosphere The air surrounding the Earth.
atmospheric pressure (barometric pressure) The pressure exerted by the atmosphere as a consequence of gravitational attraction exerted upon a column of air lying directly above the point in question.
attenuation With respect to radar, the situation in which one weather area closer to the antenna reflects or absorbs so much radar energy that more distant weather areas are effectively blocked out and not displayed.
automated surface observation system (ASOS) A self-contained weather station designed to make weather observations for aviation and other uses without operator involvement. Although similar to AWOS, ASOS measures more variables and is much more expensive.
automated weather observing station (AWOS) A self-contained weather station designed to make aviation weather observations without operator involvement.
automatic terminal information service (ATIS) The continuous broadcast of recorded noncontrol information in selected terminal areas.
Aviation Weather (AC 00-6A) A publication of the FAA Flight Standards Service and the National Weather Service containing weather knowledge essential for all pilots.
Aviation Weather Services, Advisory Circular 00-45 A publication of the FAA Flight Standards Service and the National Weather Service that is considered a supple-

ment to *Aviation Weather* and is periodically updated to reflect changes brought about by latest techniques, capabilities, and service demands.

aviation weather forecast An evaluation, according to set procedures, of those weather elements that are most important for aircraft operations. It always includes cloud height or vertical visibility, sky cover, horizontal visibility, obstructions to vision, certain atmospheric phenomena, and wind speed and direction. Complete observations include sea level pressure, temperature, dew point temperature, and altimeter setting.

backing A change in wind direction in a counterclockwise sense with respect to either space or time; opposite of veering.

banner cloud A banner-like cloud streaming off a mountain peak.

barometer An instrument for measuring the pressure of the atmosphere; the two principal types are mercurial and aneroid.

Beaufort wind scale A system of estimating and reporting wind speed, originally based on the effect of various wind speeds on the amount of canvas that a full-rigged nineteenth century frigate could carry.

blizzard A severe weather condition characterized by low temperatures and strong winds bearing a great amount of snow, either falling or picked up from the ground.

braking action A report of conditions on the airport movement area providing the pilot with a quality of braking that he might expect. Reported in terms of good, fair, poor, or nil.

breeze Wind speed 4–27 knots (4–31 mph); Beaufort scale numbers 2–6.

buildup A cloud with considerable vertical development.

Buys Ballot's law If an observer in the Northern Hemisphere stands with her back to the wind, lower pressure is to her left.

calm Wind speed less than 1 knot (1 mph); Beaufort scale 0.

cap cloud A standing or stationary caplike cloud crowning a mountain summit.

CAT *See* clear air turbulence.

cavitation The formation of a partial vacuum in a liquid by a swiftly moving solid body such as a propeller.

CAVOK Ceiling and visibility okay.

CAVU Ceiling and visibility unlimited; clear or scatter clouds and visibility greater than 10 miles.

ceiling The height ascribed to the lowest layer of clouds or obscuring phenomenon when it is reported as broken, overcast, or obscuration and not classified as thin or partial. The ceiling is termed unlimited when these conditions are not satisfied.

ceiling balloon A small balloon used to determine the height of the cloud base. The height can be computed from the ascent velocity of the balloon and the time required for its disappearance into the cloud.

ceiling classification A description or explanation of the manner in which the height of the ceiling is determined, for instance, aircraft ceiling, balloon ceiling, estimated ceiling, and measured ceiling.

ceilometer An automatic, recording cloud height indicator.

Celsius temperature scale International thermometric scale on which the freezing point of water equals 0°C and the boiling point equals 100°C at standard atmospheric pressure (760 mm Hg).

center of gravity An imaginary point where the resultant of all weight forces in a body may be considered to be concentrated for any position of the body.

center weather advisory An unscheduled weather advisory issued by a National Weather Service meteorologist assigned to an air route traffic control center for ATC use to alert pilots of existing or anticipated adverse weather conditions within the next 2 hours.

centigrade temperature scale The older name for the Celsius temperature scale; officially abandoned by international agreement in 1948, but still in common use.

chinook A warm, dry foehn wind blowing down the eastern slopes of the Rocky Mountains over the adjacent plains of the United States and Canada.

cirriform All species and varieties of cirrus, cirrocumulus, and cirrostratus clouds; descriptive of clouds composed mostly or entirely of small ice crystals, usually transparent and white with average heights ranging upward of 20,000 feet.

cirrocumulus A cirriform cloud appearing as a thin sheet of small white puffs resembling flakes or patches of cotton without shadows; sometimes confused with altocumulus.

cirrostratus A cirriform cloud appearing as a whitish veil, usually fibrous, sometimes smooth, often produces a halo phenomenon, and might totally cover the sky.

cirrus A cirriform cloud in the form of thin, white feather-like clouds in patches or narrow bands; have a fibrous and/or silky sheen; large ice crystals often trail downward a considerable vertical distance in fibrous, slanted, or irregularly curved wisps called *mares' tails*.

civil twilight The period of time before sunrise and after sunset when the sun is not more than 6° below the horizon.

clear air turbulence (CAT) Turbulence encountered by aircraft when flying through airspace devoid of clouds. Thermals and wind shear are the main causes.

clear ice (or **clear icing**) Smooth, clear, transparent ice that forms on aircraft when the temperature is close to freezing, most often in cumuliform clouds.

cloud bank A mass of clouds, usually of considerable vertical extent, stretching across the sky on the horizon, but not extending overhead.

cloud base For a given cloud or cloud layer, the lowest level in the atmosphere at which the air contains a perceptible quantity of cloud particles.

cloudburst Any sudden and heavy rain, almost always of the shower type; frequent in mountainous regions where moist air encounters orographic lifting.

cloud height The height of the cloud base above the local terrain.

cloud seeding The use of silver iodide, dry ice, or other substances to enhance precipitation.

coalescence Formation of a single water drop by the union of two or more colliding drops.

cold air mass An air mass that is colder than the underlying surface or colder relative to adjacent or surrounding air.

cold front A warm-cold air boundary with cold air advancing.

collective (or **collective pitch control**) The control used by the pilot of a helicopter to equally and simultaneously vary the pitch of all the rotor blades.

condensation trail (or **contrail**) A cloud-like streamer frequently observed to form behind aircraft flying in clear, cold, humid air.

contact approach An approach wherein an aircraft on an IFR flight plan, having an air traffic control authorization, operating clear of clouds with at least 1 mile flight visibility and a reasonable expectation of continuing to the destination airport in those conditions, may deviate from the instrument approach procedure and proceed to the destination airport by visual reference to the surface. This approach will only be authorized when requested by the pilot and the reported ground visibility at the destination airport is at least 1 statute mile.

continental air mass An air mass that forms overland, making it generally dry. It can be warm or cold.

convection The circulatory motion that occurs in a fluid at a nonuniform temperature owing to the variation of its density and the action of gravity; also, the transfer of heat by this automatic circulation of the fluid. In meteorology, atmospheric motions that are predominantly vertical, resulting in vertical transport and mixing of atmospheric properties.

convective sigmet An advisory of concern to all aircraft that is issued hourly during periods of hazardous convective weather, such as tornadoes, thunderstorms, and hail. Convective sigmets are issued by the National Severe Storms Forecast Center in Kansas City.

convective storms Storms created by rising warm air: thunderstorms.

convergence The condition that exists when the distribution of winds within a given area results in a net horizontal inflow of air into the area.

Coriolis force A deflecting force acting on a body in motion and resulting from the earth's rotation. It deflects air currents to the right in the Northern Hemisphere and to the left in the Southern Hemisphere, having an effect on wind direction.

crab angle A wind correction angle. The angular difference between the course and heading of an aircraft due to the effects of a crosswind.

crosswind A wind blowing in a direction perpendicular to the course of a moving object.

cumuliform A general term descriptive of all clouds exhibiting vertical development in contrast to horizontally extended stratiform types.

cumulonimbus A cumuliform cloud type that is heavy and dense, with considerable vertical extent in the form of a massive tower, often with tops in the shape of an anvil or massive plume. Underneath the base of cumulonimbus, which is often very dark, there frequently exists virga, precipitation, and low ragged clouds (*scud*), perhaps merged with the cloud formation. A cumulonimbus frequently has lightning and thunder, and sometimes hail. It occasionally produces a tornado over land or a waterspout over water. A cumulonimbus is the ultimate manifestation of the growth of a cumulus cloud, occasionally extending well into the stratosphere.

cumulus A cloud in the form of individual detached domes or towers that are usually dense and well defined. It develops vertically in the form of rising mounds; the bulging upper part often resembles a cauliflower. Sunlit parts of these clouds are mostly brilliant white and the bases are relatively dark and nearly horizontal.

cyclic (or *cyclic pitch control*) The control by which the pilot of a helicopter changes the pitch of the rotor blades individually during a cycle of revolution to control the tilt of the rotor disc, which affects the direction and velocity of horizontal flight.

cyclone An area of low atmospheric pressure that has a closed circulation that is cyclonic (counterclockwise in the Northern Hemisphere and clockwise in the Southern Hemisphere).

deepening A decrease in the central pressure of a pressure system; usually applied to a low rather than to a high.

density altitude Pressure altitude corrected for temperature and humidity. (*Pressure altitude* is the indication of an altimeter when set to 29.92, typically on the ground to determine density altitude for preplanning takeoff performance.)

depression In meteorology, an area of low pressure: a *low* or *trough*.

dew Water condensed onto objects at or near the ground, due to the fact that their temperatures have fallen below the dew point temperature of the surrounding air, but not below freezing.

dew point (or **dew point temperature**) The temperature to which a sample of air must be cooled, while the mixing ratio and barometric pressure remain constant, in order to attain saturation by water vapor.

direct user access terminal (DUAT) A computerized weather briefing service under the auspices of FAA, presently provided free of charge by two independent weather vendors to all certificated pilots in the United States.

distance measuring equipment (DME) A navigation aid that provides a pilot with the straightline distance in nautical miles from the selected ground station.

distant early warning identification zone (DEWIZ) An ADIZ over the coastal waters of Alaska.

diurnal An event that occurs every 24 hours, especially pertaining to a cycle completed within a 24-hour period.

divergence The condition that exists when the distribution of winds within a given area results in a net horizontal flow of air outward from the region. In divergence at lower levels, the resulting deficit is compensated for by a downward movement of air from aloft.

Doppler radar Radar that measures the speed and direction of a moving object, such as dust or water droplets suspended by wind.

downburst High-velocity, short-lived, vertical air movement that is typically released from a thunderstorm or thundershower. A downburst can cause a pilot to lose control of an aircraft.

downdraft A relatively small-scale, downward moving current of air that does not dramatically affect control of an aircraft.

drag That component of the resultant of the aerodynamic forces acting on a body that acts parallel to the relative airflow; the force that tends to resist movement of an airfoil through the air.

drizzle Very small precipitation drops that appear to float with air currents while falling in an irregular path.

dust storm An unusual, frequently severe weather condition characterized by strong winds and dust-filled air over an extensive area.

eddy A local irregularity of wind in a larger-scale wind flow. Small-scale eddies produce turbulent conditions.

electronic flight instrument system (EFIS) Video display tubes that replace electromechanical gauges: airspeed indicator, altimeter, compass, navigation instruments, engine instruments, and the like. The EFIS screens provide virtually unlimited possibilities for presenting flight data. *See also* glass cockpit.

emergency landing A landing made because of a technical or other serious problem. *See* forced landing and precautionary landing.

en route flight advisory service (EFAS) A service specifically designed to provide, upon pilot request, timely weather information pertinent to the type of flight, intended route, and altitude. The common EFAS frequency for aircraft at altitudes below 18,000 feet is 122.0 MHz. FSSs providing this service are listed in the *A/FD*. *See* flight watch.

ergonomics An applied technology that in part studies the physical interrelation of people and devices; also called *human factors engineering*.

estimated ceiling A ceiling classification applied when the ceiling height has been estimated by the observer or has been estimated by some other method.

eye The roughly circular area of calm or relatively light winds and comparatively fair weather at the center of a well-developed tropical cyclone.

Fahrenheit temperature scale A thermometric scale on which the freezing point of water equals 32°F and the boiling point equals 212°F at standard atmospheric pressure (29.92).

fall wind A cold wind blowing downslope. Fall wind differs from foehn in that the air is initially cold enough to remain relatively cold despite compressional heating during descent.

Federal Aviation Administration (FAA) The federal agency that is responsible for the safety, regulation, and promotion of civil aviation, and the safe and orderly use of the National Airspace System.

fixed-base operator (FBO) Typically, the aviation equivalent of a combined automobile service station, auto maintenance shop, new- and used-car dealership, and student driving school.

field elevation The officially designated elevation of an airport above mean sea level, taken as the highest point on any of the runways of the airport.

figmo An acronym that abbreviates "forget it, got my orders," which is used in the military to describe the "I-don't-care" attitude of someone who has received an assignment to another unit and is unconcerned about job performance in the present unit.

filling An increase in the central pressure of a pressure system, which is the opposite of deepening pressure. Filling is more commonly applied to a low rather than a high.

flicker vertigo A state of spatial confusion brought on by misleading sensory information transmitted to the brain, such as that caused by the rhythmic flickering of sunlight through a propeller.

flight following *See* traffic advisories.

flight forecast An aviation weather forecast for a specific flight.

flight service station (FSS) Air traffic facilities that provide pilot briefing, en route communications, and VFR search and rescue services, assist lost aircraft and aircraft in emergency situations, relay ATC clearances, originate NOTAMs, broadcast aviation weather and National Airspace System information, receive and process IFR flight plans, and monitor navigational aids.

flight watch A shortened term for use during air-to-ground contacts to identify an FSS that provides the en route flight advisory service, which is fully defined in this glossary.

foehn A warm, dry wind on the lee side of a mountain range. The warmth and dryness is due to adiabatic compression upon descent.

fog An atmospheric condition consisting of numerous minute water droplets suspended in the air near the surface—a cloud that is on or very close to the ground.

forced landing A landing made when conditions dictate that it is no longer possible to remain airborne, such as after an engine failure in a single-engine airplane.

forward-looking infrared (FLIR) A vision-enhancing system that presents images based on the detection of radiation in the infrared range (heat) rather than in the visual range; a night-vision device that is typically mounted on military helicopters.

fractus Clouds in the form of irregular shreds that appear to be torn and have a clearly ragged appearance. The term only applies to stratus and cumulus clouds: cumulus fractus and stratus fractus.

freezing level Level in the atmosphere at which the temperature is 0°C or 32°F.

freezing nucleus Particle on which the freezing of water occurs.

freezing point Temperature of solidification of a liquid under given conditions.

freezing rain Supercooled raindrops that turn to ice when they come in contact with something.

front Boundary between air masses of different densities and usually different temperatures.

frost Ice crystal deposits formed by sublimation (conversion of water directly to ice) when temperature and dew point are below freezing.

funnel cloud A rotating column of air extending from a cloud, but not reaching the ground.

gale Wind speed of 28–55 knots (32–63 mph); Beaufort scale numbers 7–10.

geostrophic wind That horizontal wind velocity at which the Coriolis acceleration exactly balances the horizontal pressure force; it is directed along contour lines or isobars.

glass cockpit An aircraft cockpit dominated by electronic flight instrument systems (EFIS).

glaze A coating of ice, generally clear and smooth, formed by the freezing of supercooled water on a surface.

global positioning system (GPS) A navigation system that utilizes satellites to determine the position of objects on the earth's surface or in the air; also called *navstar*.

ground clutter Pertaining to radar, a cluster of echoes, generally at short range, reflected from ground targets.

ground fog A layer of fog, often less than 200 feet high, that forms when the ground cools.

gust A sudden brief increase in the speed of the wind, followed by a lull or slackening, with variations of 10 knots or more between peaks and lulls.

gust front Wind flowing out from a thunderstorm.

hail Precipitation composed of balls or irregular lumps of ice with diameters between 5 and 50 millimeters.

halo A prismatically colored or whitish circle, or arcs of a circle, with the sun or moon at its center.

haze A type of lithometeor composed of fine dust or salt particles dispersed through a portion of the atmosphere. The particles are so small that neither can they be felt on the skin nor seen with the eye, but the collective presence diminishes visibility.

headwind A wind moving in a direction opposite to the heading of a moving object, thus opposing the object's intended progress.

heat lightning Glowing flash in clouds. No thunder is heard because heat lightning is too far away.

high An area of high barometric pressure, with its attendant system of winds; an anticyclone, which is fully defined in this glossary.

human engineering *See* ergonomics.

humidity Water vapor content of the air; may be expressed as absolute humidity, specific humidity, relative humidity, or mixing ratio.

hurricane A tropical cyclone in the Western Hemisphere with winds in excess of 64 knots (73 mph).

hurricane-force wind Wind speed that is faster than 64 knots (73 mph); Beaufort numbers 12–17.

hydrology Science that deals with the occurrence, circulation, distribution, and properties of the waters of the earth and its atmosphere.

hydrometeor A general term for atmospheric water in any of its forms, such as rain, ice crystals, hail, fog, or clouds.

hygrometer Instrument used to measure the water vapor content of the air.

hypoxia A deficiency in the amount of oxygen that reaches the body's tissues. The deficiency can be caused by flight at high altitude.

ice crystals A type of precipitation composed of unbranched crystals in the form of needles, columns, or plates. They usually have a very slight downward motion and might fall from a cloudless sky.

ice fog A type of fog that is composed of minute suspended particles of ice and usually occurs at very low temperatures.

ice pellets Falling drops of frozen water, also called *sleet*.

icing In general, any deposit of ice that is formed on an object.

indefinite ceiling A ceiling classification denoting vertical visibility into a surface-based obscuration.

instrument flight rules (IFR) Rules governing the procedures for conducting instrument flight.

instrument landing system (ILS) A precision approach system that provides a pilot with horizontal and vertical guidance toward the end of a runway for landing.

instrument meteorological conditions (IMC) Weather and/or flight conditions to which instrument flight rules apply.

inversion An increase in temperature with height, which is a reversal of the normal decrease in temperature with height in the troposphere.

isobar A line of equal or constant barometric pressure.

isotherm Line connecting points on a map that have the same temperature.

jet stream A narrow band of upper-atmosphere wind with speeds greater than 57 mph.

katabatic wind Any wind blowing down an incline. If warm, it is a foehn; if cold, it might be a fall wind or a gravity wind.

Kollsman window The altimeter setting window of barometric altimeters that gives the pilot a way to adjust the altimeter for atmospheric pressure variations.

knot A unit of speed equal to 1 nautical mile per hour.

land breeze Winds blowing from any body of water toward land, caused by the temperature difference when the sea surface is warmer than the land surface.

layer In reference to sky cover, clouds, or other obscuring phenomena with bases that are approximately at the same level.

lapse rate Rate of decrease of temperature with increasing height.

lee wave A wave disturbance in airflow due to some barrier in the flow, for instance, a hill or mountain.

lenticular clouds Lens-shaped clouds that form on the mountain summits indicating the presence of moist air flowing across the terrain.

lift That component of the resultant of the aerodynamic forces acting on a body that acts at right angles to the relative airflow. Lift is the force derived from an airfoil, according to Bernoulli's Principle: *venturi effect*.

light air Wind speed 1–3 knots (1–3 mph); Beaufort scale 1.

lightning A visible discharge of static electricity produced by a thunderstorm.

lithometeor The general term for dry particles suspended in the atmosphere, such as dust, haze, smoke, and sand.

local winds Winds that, over a small area, differ from those that would be appropriate to the general pressure distribution.

loran-C An area navigation system that computes position based on signals received from a primary and at least two secondary transmitting stations.

low An area of low barometric pressure, with its attendant system of winds. It is also called a depression or cyclone.

manifold pressure The air pressure measured at a certain point in the induction manifold of a reciprocating engine.

maritime air mass An air mass that forms over an ocean, making it humid. It might be warm or cold.

mean sea level (MSL) The average height of the surface of the sea for all stages of tide. It is a reference for ground elevations throughout the United States, and it is also a reference for aircraft altitude.

measured ceiling A ceiling classification applied when the ceiling value has been determined by instruments or the known heights of unobscured portions of objects, other than natural landmarks.

melting level Level at which ice crystals and snowflakes melt in the course of their descent through the atmosphere.

melting point Temperature that change occurs from a solid to the liquid phase of a substance. It is a function of pressure.

mercurial barometer A barometer in which pressure is determined by balancing air pressure against the weight of a column of mercury in an evacuated glass tube.

mesostorm A formation of several individual air mass thunderstorms into a single, highly organized storm system that can cover 40,000 square miles and last for up to 12 hours.

meteor Any natural phenomenon in the atmosphere.

meteorology The science of the atmosphere.

microburst A downburst less than 2.5 miles in diameter.

millibar A unit of pressure that directly expresses the force exerted by the atmosphere; equal to 1,000 dynes per cubic meter or 100 pascals; convenient for reporting atmospheric pressure.

mist Drizzle or heavy fog.

mixed ice (or **mixed icing**) Aircraft icing that is a mixture of clear and rime ice.

mixing ratio In a system of moist air, the dimensionless ratio of the mass of water vapor to the mass of dry air.

monsoon A wind that in summer blows from sea to a continental interior, bringing copious rain, and in winter blows from the interior to the sea, resulting in sustained dry weather.

mountain waves A standing or lee wave to the lee of a mountain barrier.

mountain winds Winds that blow either up or down mountains, caused by different rates of heating and cooling or mountaintops and valleys.

moving map A type of navigational equipment that can present aeronautical chart information on a video tube or other display while the aircraft is in flight.

moving target indicator (MTI) A function selection on air traffic control radar that is used to decrease ground clutter by only detecting objects in motion.

multicell storms Thunderstorms consisting of clusters of single-cell thunderstorms.

National Meteorological Center (NMC) The National Weather Service office in Coral Gables, Florida, that tracks and forecasts hurricanes and other weather in the Atlantic, Gulf of Mexico, Caribbean Sea, and parts of the Pacific.

National Severe Storms Forecast Center The National Weather Service facility in Kansas City, Missouri, that issues watches for severe thunderstorms and tornadoes across the nation.

National Severe Storms Laboratory (NSSL) The National Oceanic and Atmospheric Administration laboratory in Norman, Oklahoma, that studies severe weather, such as thunderstorms and the damaging winds, hail, and tornadoes that can be attributed to them.

National Transportation Safety Board (NTSB) The federal agency tasked with investigating aircraft and other transportation accidents, especially fatality accidents. (The FAA might be the sole investigative agency for selected nonfatality aviation accidents.)

National Weather Service (NWS) The federal agency that observes and forecasts weather. It is part of the National Oceanic and Atmospheric Administration (NOAA), which is part of the Department of Commerce.

nautical mile One minute of arc on a meridian (one minute of latitude); one minute of arc on the earth's equator. One nautical mile equals approximately 1.15 statute miles or 1.85 kilometers; therefore, to convert knots to miles per hour simply multiply knots by 1.15. To convert knots to kilometers per hour, just double knots for a rough approximation, or multiply the knots by 1.85, to be more precise.

NAVSTAR *See* global positioning system.

nephology The study of clouds.

NEXRAD Next generation weather radar system being installed in the 1990s by the National Weather Service, the Department of Defense, and the Federal Aviation Administration; officially designated WSR-88D, for *W*eather *S*ervice *R*adar developed in 19*88*, *D*oppler.

nimbostratus A principal cloud type, gray-colored, often dark, the appearance of which is rendered diffuse by more or less continuously falling rain or snow, which in most cases reaches the ground. It is thick enough to blot out the sun.

nocturnal Occurring during the hours between sunset and sunrise.

nowcasting A form of very short-range weather forecasting of current weather along with forecasts of up to about 2 hours.

numerical weather prediction Forecasting by digital computers solving mathematical equations, used extensively in weather services throughout the world.

obscuration Denotes sky hidden by a surface-based obscuring phenomenon. Vertical visibility is restricted.

obscuring phenomenon An atmospheric phenomenon, other than clouds, that obscures a portion of the sky at the point of observation.

occluded front (or **occlusion**) A boundary between cool, cold, and warm air masses. A composite of two fronts as a cold front overtakes a warm front.

operational weather limits The limiting values of ceiling, visibility and wind, or runway visual range, that are established as safety minima for aircraft landings and takeoffs.

orography A branch of physical geography that deals with mountains.

orographic precipitation Precipitation caused by the ascent of moist air over a mountain barrier.

orographic turbulence Turbulence caused by the movement of air over mountains.

ozone An unstable form of oxygen. The heaviest concentrations are in the stratosphere. It absorbs most ultraviolet solar radiation. Ozone is corrosive to some metals.

parcel A volume of air that is small enough to contain uniform distribution of its meteorological properties and large enough that it will remain relatively self-contained and respond to all meteorological processes.

pascal Name given to the unit of pressure in the International System of Units (SI); equal to one newton per square meter or 0.01 millibar.

partial obscuration A designation of sky cover when part of the sky is hidden by a surface-based obscuring phenomenon.

pilot weather report (PIREP) A radio report from a pilot regarding meteorological conditions encountered in flight.

PJ An Air Force pararescue specialist.

plow wind The spreading downdraft of a thunderstorm; a strong, straight-line wind in advance of the storm.

point of equal time (PET) The time when an aircraft that is flying between two points will be exactly halfway between the two points with respect to time, but not necessarily distance, taking into account present or expected wind conditions.

point of no return (PNR) The time when an aircraft, which is flying between two points, will not have enough fuel on board to return to the point of departure, taking into account present or expected wind conditions.

precautionary landing An emergency landing that is not immediately required, but that is made to avoid the possibility of worsening conditions, due to technical, meteorological, or other conditions.

precipitable water The amount of water, expressed as a depth or as a mass, that would be obtained if all the water vapor in a specified column of the atmosphere were condensed and precipitated.

precipitation Any and all forms of water particles, liquid or solid, that fall from the atmosphere and reach the surface.

precipitation fog Fog that forms when precipitation falls into cold air.

prefrontal squall lines Lines of thunderstorms ahead of an advancing cold front.

pressure altimeter An aneroid barometer with a scale graduated in altitude instead of pressure units.

pressure gradient The rate of decrease of pressure per unit distance in a fixed time.

pressure jump A sudden, significant increase in station pressure.

prevailing visibility In the United States, the greatest horizontal visibility that is equaled or exceeded throughout half of the horizon circle.

prevailing wind The direction that the wind blows from most frequently.

prognosis A presentation of forecast weather conditions as they are expected to exist at some time or interval of time.

prognostic chart A chart of expected or forecast conditions.

pulsating visual slope indicators (PLASI) A type of visual approach slope indicator for guidance to a landing. *See* VASI.

QFE Atmospheric pressure at field elevation. Setting QFE in the Kollsman window of an aircraft barometric altimeter will show zero feet when the aircraft is at field elevation.

QNH Same as altimeter setting, which is defined in this glossary. Setting QNH in the Kollsman window of an aircraft barometric altimeter will show the field elevation when the aircraft is at field elevation.

radar An electronic instrument used for the detection and ranging of distant objects of such composition that they scatter or reflect radio energy. Because hydrometeors can scatter radio energy, weather radars operating on certain frequency bands can detect the presence of precipitation or clouds, or both. Radar is the acronym of *radio detection and ranging*.

radiation fog Fog characteristically resulting when radiational cooling of the earth's surface cools the air near the ground to or below its initial dewpoint on calm, clear nights.

radiosonde A balloon-borne instrument for the simultaneous measurement and transmission of meteorological data. It includes transducers for the measurement of pressure, temperature, and humidity, a modulator, a switching mechanism, and a radio transmitter.

radiosonde balloon A balloon used to carry a radiosonde aloft.

radiosonde observation An evaluation of upper air temperature, pressure, and humidity from radio signals received from a balloon-borne radiosonde.

rain Precipitation composed of liquid water drops more than 0.5 millimeter in diameter, falling in relatively straight, but not necessarily vertical, paths.

rawin A method of winds aloft observation accomplished by radar tracking of a balloon-borne target or radiosonde.

rawinsonde observation A combined winds aloft and radiosonde observation. Winds are determined by tracking the radiosonde by radio direction finder or radar.

relative humidity The ratio of the existing amount of water vapor in the air at a given temperature to the maximum amount that could exist at that temperature. The ratio is usually expressed as a percentage: 27 percent relative humidity.

relative wind or **airflow** The velocity of the air with reference to a body in it. The direction of airflow with respect to an airfoil.

ridge An elongated area of relatively high pressure; usually running north-south and associated with, and most clearly identified as, an area of maximum anticyclonic curvature of the wind flow.

rime An accumulation of granular ice tufts on the windward side of exposed objects that is formed from supercooled fog or cloud and built out directly against the wind.

rime ice (or **rime icing**) The formation of a white or milky and opaque granular deposit of ice formed by the rapid freezing of super-cooled water drops as they impinge upon a surface, such as the leading edge of a wing.

rocket sonde (or **meteorological rocket**) A rocket designed primarily for routine upper air observations in the lower 250,000 feet of the atmosphere, especially that portion inaccessible to balloons (above 100,000 feet).

roll cloud A dense and horizontal roll-shaped accessory cloud that is located on the lower leading edge of a cumulonimbus cloud. Less often it might be associated with a rapidly developing cumulus cloud. The roll cloud indicates turbulence, sometimes severe, that should be avoided; do not land or take off when a roll cloud is visible in the vicinity of the airport. (A roll cloud is sometimes improperly called a rotor cloud, the next defined term.)

rotor cloud A turbulent altocumulus cloud formation found in the lee of some large mountain barriers, the air in the cloud rotates around an axis parallel to the range; indicative of severe to extreme turbulence. (A rotor cloud is sometimes improperly called a roll cloud, the previously defined term.)

RPM Revolutions per minute

runway temperature The temperature of the air just above a runway, ideally at engine and/or wing height, used to determine density altitude.

runway visibility The meteorological visibility along an identified runway determined from a specified point on the runway. It might be determined by a transmissometer or by an observer.

runway visual range (RVR) The horizontal distance a pilot will see down the runway from the approach end; derived by transmissometer.

Saint Elmo's fire The slow discharge of natural electrical currents that appears as a blue corona on grounded metal objects, chimney tops, ship masts, and aircraft protrusions during stormy weather.

saturation The condition of the atmosphere when the amount of water vapor present is the maximum possible at the existing temperature.

scud Small detached masses of stratus fractus clouds below a layer of higher clouds; the higher clouds are usually nimbostratus.

scud running The practice of flying below low layers of clouds, often with little forward visibility.

sea breeze Winds blowing inland from any body of water that are caused by the temperature difference when the land surface is warmer than the sea surface.

sea fog *See* advection fog.

sea level pressure The atmospheric pressure at mean sea level that is used as a common reference for analyses of surface pressure patterns. Sea level pressure is either directly measured by stations at sea level or empirically determined from the station pressure and temperature by stations not at sea level.

search-and-rescue (SAR) A service to seek missing aircraft and to help aircraft that need assistance.

seat-of-the-pants Flying by sight, touch, and feeling. (Ideally, the wind is whistling around your goggles and ears.)

severe thunderstorm A thunderstorm with winds faster than 57 mph or hailstones three-quarters of an inch, or larger, in diameter.

shower Precipitation from a cumuliform cloud, characterized by the suddenness of beginning and ending, by the rapid change in intensity, and usually by a rapid change in the condition of the sky. The solid or liquid water particles are usually bigger than the corresponding elements in other types of precipitation.

significant meteorological information (SIGMET or sigmet) A weather advisory about potentially hazardous weather conditions of concern to all aircraft, such as icing, turbulence, dust storms, sandstorms, and the like, that is issued when necessary.

sky cover The amount of sky covered or concealed by clouds or an obscuring phenomenon. Sky cover is reported in tenths, supplemented with a descriptive term: clear, scattered, broken, overcast, partial obscuration, or obscuration.

slant range The line-of-sight distance between two objects that are not at the same elevation.

slant visibility The distance at which an airborne observer can see and distinguish objects on the ground.

sleet Frozen or partly frozen rain.

smog Fog contaminated by industrial pollutants; a mixture of smoke and fog.

snow Precipitation composed of white or translucent ice crystals, chiefly in complex branched hexagonal forms.

snow flurry Snow shower, particularly of a very light and brief nature.

snow grains Precipitation of very small, white opaque particles of ice, fairly flat or elongated, with diameters less than one millimeter.

snow pellets (or **soft hail**) Precipitation of white, opaque, spherical, or conical ice particles that are crisp and easily crushed and that have diameters of 2–5 millimeters.

solar radiation The total electromagnetic radiation emitted by the sun.

sounding In meteorology, an upper-air observation—a radiosonde observation.

source region An extensive area of the earth's surface characterized by relatively uniform conditions where large masses of air remain long enough to take on characteristic temperature and moisture properties imparted by that surface.

special VFR conditions Weather conditions that are less than basic VFR weather conditions and that permit flight under VFR in certain airspace. (*See* the *Airman's Information Manual.*)

specific humidity In a system of moist air, the dimensionless ratio of the mass of water vapor to the total mass of the system.

squall A line of thunderstorms, generally continuous, across the horizon. The squall line is associated with prefrontal activity.

stall The loss of streamlined airflow over an airfoil caused by excessive angle of attack that results in loss of lift.

standard atmosphere Atmospheric conditions in which (1) the air is a dry, perfect gas; (2) the temperature at sea level is 59° F/15° C; (3) the pressure at sea level is 29.9213 inches/ 1013.250 millibars of mercury (Hg)/760 mm Hg; and (4) the temperature gradient is approximately 3.5°F/2°C per 1,000-foot change in altitude.

standing wave A wave that remains stationary in a moving fluid. In aviation, it is used most commonly to refer to a lee wave or mountain wave.

stationary front A warm-cold boundary with neither cold nor warm air advancing.

station pressure The atmospheric pressure computed using station elevation as the reference datum level. Station pressure is usually the base value from which sea level pressure and altimeter setting are determined.

steam fog Fog formed when cold air moves over relatively warm water or ground.

stratiform Clouds with extensive horizontal development, as contrasted to vertically developed cumuliform clouds. Stratiform clouds are characteristic of stable air and are composed of small water droplets.

stratocumulus A low cloud, predominantly stratiform in gray and/or whitish patches or layers that might merge.

stratosphere The layer of the atmosphere 7–30 miles above the surface.

stratus A low, gray cloud layer or sheet with a fairly uniform base that sometimes appears in ragged patches. Stratus clouds seldom produce precipitation, but might produce drizzle or snow grains.

sublimation Direct phase changes of water: vapor to ice or ice to vapor.

supercell A severe thunderstorm that usually lasts several hours, often spinning out a series of strong tornadoes.

supercooled water Water at temperatures colder than freezing.

surface inversion An inversion with its base at the surface, often caused by cooling of the air near the surface as a result of terrestrial radiation, especially at night.

surface visibility Visibility observed from eye-level above the ground.

switchology The manner that switches and instruments are designed. In the context of human engineering, switchology is important because switches and instruments that are poorly designed or improperly placed might cause a person to operate the switches or interpret the instruments incorrectly.

synoptic A general view of the whole, for instance a synoptic weather map or synoptic weather situation in which the major weather systems over a large geographical area are depicted or discussed.

tailwind A wind blowing in the same direction as the heading of a moving object, assisting the object's intended progress.

temperature In general, the degree of hotness or coldness as measured by the Celsius or Fahrenheit temperature scale on a thermometer.

temperature inversion Vertical temperature distribution such that temperature increases with height.

terminal Doppler weather radar (TWDR) Doppler radars that are being installed at major airports to detect microbursts.

thermometer Instrument used in the measurement of temperature.

thunder Sound produced by a lightning discharge.

thunderstorm A local storm produced by a cumulonimbus cloud that is always accompanied by lightning and thunder.

tornado A strong, rotating column of air extending from the base of a cumulonimbus cloud to the ground, and nearly always observed as "funnel-shaped." A tornado is the most destructive small-scale atmospheric phenomenon.

towering cumulus A rapidly growing cumulus in which height exceeds width.

tower visibility Prevailing visibility determined from the control tower.

transmissometer A device system that determines visibility by measuring the transmission of light through the atmosphere. It is the measurement source for determining runway visual range and runway visibility value.

traffic advisories Advisories issued to alert pilots to other known or observed air traffic that might be in such proximity to the position or intended route of flight of their aircraft to warrant their attention.

transcribed weather broadcast (TWEB) A continuous recording of meteorological and aeronautical information that is broadcast on VOR facilities for pilots.

tropical cyclone A low-pressure system in which the central core is warmer than the surrounding atmosphere. This is a general term for a cyclone that originates over tropical oceans.

tropical depression A tropical cyclone with maximum sustained winds less than 39 mph near the surface.

tropical storm A tropical cyclone with 39–74 mph winds.

tropopause The boundary between the troposphere and the stratosphere, usually characterized by an abrupt change in lapse rate. Its height varies from 35,000–65,000 feet. Regions above the tropopause have better atmospheric stability than regions below.

troposphere That portion of the atmosphere from the earth's surface to the tropopause, the lowest 35,000–65,000 feet of the atmosphere characterized by decreasing temperature with height, and by appreciable water vapor.

trough An elongated area of relatively low atmospheric pressure, usually associated with and most clearly identified as an area of maximum cyclonic curvature of the wind flow.

true wind direction The direction from which the wind is blowing, with respect to true north.

turbulence Irregular motion of the atmosphere produced when air flows over a comparatively uneven surface, such as the surface of the earth, or when two currents of air flow past or over each other in different directions or at different speeds.

twilight The intervals of incomplete darkness following sunset and preceding sunrise. The time at which evening twilight ends or morning twilight begins is determined by arbitrary convention. (*See* civil twilight.)

typhoon A tropical cyclone with winds 75 mph or more in the North Pacific, west of the International Dateline.

uncontrolled airport An airport without an operating control tower.

undercast A cloud layer of ten-tenths (1.0) coverage as viewed from an observation point above the layer

unicom A nongovernment communication facility that can provide airport information at certain airports. The frequency 123.0 MHz is used at airports served by an air traffic control tower or FSS and 122.8 MHz is used for other landing areas.

unlimited ceiling A clear sky or a sky cover that does not meet the criteria for a ceiling.

updraft A relatively small-scale current of air that is moving upward.

upper air That portion of the atmosphere that is above the lower troposphere.

upper air observation A measurement of atmospheric conditions aloft, above the effective range of a surface weather observation. Elements evaluated include temperature, humidity, pressure, wind speed, and wind direction.

upslope fog Fog that forms in humid air flowing uphill over rising terrain and is adiabatically cooled to or less than its initial dew point.

UTC Coordinated universal time, formerly officially called GMT, but can be heard informally referred to as Zulu or Greenwich Meantime. The time of day based on a 24-hour atomic clock.

veering Shifting of the wind in a clockwise direction with respect to either space or time. It is the opposite of backing.

venturi effect The principle that describes what happens when a quantity of air moves through a restricted space causing the velocity of the air to increase and its pressure to decrease; also called the *Bernoulli principle*.

vertical visibility The distance that an observer can see vertically into a surface-based obscuring phenomenon such as fog, rain, or snow. The distance estimate must be based upon ceiling balloon ascensions or ceiling light projector measurements.

very-high-frequency omnidirectional range (VOR) A navigation aid that generates directional information and transmits it by ground equipment to an aircraft, providing 360 magnetic courses to and from the station. VOR is presently the primary navigation system for civil aviation in the United States and most other countries.

VFR conditions Weather conditions equal to or better than the minimum for flight VFR.

VFR not recommended An advisory provided by an FSS briefer to a pilot during a preflight or in-flight weather briefing that flight under VFR is not recommended, considering the current and/or forecast weather conditions at or below VFR minimums.

VFR-on-top An ATC authorization for an IFR aircraft to operate in VFR conditions at any appropriate VFR altitude.

VFR-over-the-top The operation of an aircraft over-the-top of clouds under VFR when it is not being operated on an IFR flight plan.

virga Precipitation falling from a cloud, usually in wisps or streaks, and evaporating before it reaches the ground.

visual approach slope indicator (VASI) An airport lighting facility providing vertical visual approach guidance to aircraft during approach to landing.

visual flight rules (VFR) Rules that govern the procedures for conducting flight under visual conditions.

visibility The greatest distance at which it is just possible to see and recognize with the unaided eye, in the daytime, a prominent dark object against the sky at the horizon, and at night, a known preferably unfocused, moderately intense light source.

visual approach An approach wherein an aircraft on an IFR flight plan, operating in VFR conditions under the control of an air traffic control facility and having an air traffic control authorization, may proceed to the airport of destination in VFR conditions.

visual meteorological conditions (VMC) Relatively favorable weather and/or flight conditions to which visual flight rules apply.

$\mathbf{V_{NE}}$ Never-exceed speed

$\mathbf{V_{NO}}$ Normal operating speed

$\mathbf{V_{Y}}$ Speed for the best rate of climb

wake turbulence A phenomenon that results from the passage of an aircraft through the atmosphere. The term includes vortices, thrust-stream turbulence, jet blast, jet wash, propeller wash, and rotor wash on the ground and in the air.

wall cloud The well-defined bank of vertically developed clouds with a wall-like appearance that form the outer boundary of the eye of a well-developed tropical cyclone.

warm air mass An air mass that is warmer than the underlying surface or warmer relative to adjacent or surrounding air.

warm front Any nonoccluded front that moves in such a way that warmer air replaces colder air.

waterspout A tornado or weaker vortex from the bottom of a cloud to the surface of a body of water.

water vapor Water in the invisible gaseous form.

weather The state of the six meteorological elements in the atmosphere: air temperature, humidity, clouds, precipitation, atmospheric pressure, and wind.

weather advisory In aviation weather forecast practice, an expression of anticipated hazardous weather conditions as they affect the operation of air traffic and as prepared by the National Weather Service.

weather forecast A statement or projection of the future state of the atmosphere with specific reference to one or more associated weather elements.

weather radar Radar specifically designed for observing weather.

wet-bulb temperature The lowest temperature that can be obtained on a wet-bulb thermometer in any given sample of air. It is used to compute dew point and relative humidity.

wet-bulb thermometer A thermometer with a moistened muslin-covered bulb to measure wet-bulb temperature.

wind Air in motion relative to the surface of the earth; almost exclusively used to denote the horizontal component.

wind direction The direction from which the wind is blowing, measured in points of the compass or in azimuth degrees.

wind rose A flower-like diagram indicating the relative frequencies of different wind directions for a given station and period of time.

wind shear The rate of change of wind velocity per unit distance. It is conventionally expressed as vertical or horizontal wind shear.

wind sock A fabric cone attached to a metal ring and used to indicate wind direction.

wind speed Rate of wind movement in distance per unit time.

wind vane A device that indicates wind direction.

wind vector An arrow representing wind velocity, drawn to point in the direction of the wind and with a length proportional to wind speed.

wind velocity A vector term that includes wind speed and wind direction.

windchill factor The apparent temperature felt on human skin owing to the combination of temperature and wind speed.

winds aloft The wind speed and direction at certain levels in the atmosphere above the level reached by surface weather observations.

winds-aloft observation The measurement and computation of wind speeds and directions at certain levels above the surface of the earth.

zephyr Any soft, gentle breeze.

Resources

These research references for *Flying in Adverse Conditions*, plus other references mentioned within the text, would be worthwhile in every pilot's library.

"Cold Facts on Ice," by William D. Waldock, *Aviation Safety*, Jan. 1, 1993.

"Why Icing Forecasts Aren't Reliable," by Tom Miner, *IFR*, November 1990.

Planet Earth—Storm, by A.B.C. Whipple, Time-Life Books, 1982.

Aviation Weather for Pilots and Flight Operations Personnel, AC 00-6A, Federal Aviation Administration, Revised 1975.

Avoiding Common Pilot Errors: An Air Traffic Controller's View, by John Stewart, TAB/McGraw-Hill, Inc., 1989.

Better Homes and Gardens Family Medical Guide.

Aviation Weather, Forces to be Reckoned With, by Richard L. Taylor, 1991

Index

Illustration page numbers are in boldface

M

N

O

P

U

V

W

About the author

R. Randall Padfield has combined a career in writing and flying for more than 20 years. His first published article was in *Air Rescue* in 1974, which he wrote while serving as a search and rescue pilot assigned to an air force squadron in Keflavik, Iceland. Padfield's articles and stories have since been published in numerous publications, including *Aviation International News*, *Rotor & Wing International*, *Professional Pilot*, *Vertiflite*, *Helicopter World*, *Panorama*, *Fliegermagazin*, *Viten*, *The Toastmaster*, *The Norseman*, *Stars and Stripes*, *The MAC Flyer*, and *The SAGA Weekly Post*.

A 1971 U.S. Air Force Academy graduate, Padfield has more than 9,000 flight hours and airline transport certificates in both airplanes and helicopters. He left the Air Force in 1977 and worked for Helikopter Service of Norway, a North Sea helicopter operator, for nearly 12 years, flying IFR-instrumented twin-engine helicopters. He also worked as a flight simulator and ground instructor with that company.

During this time, Padfield also wrote training manuals, briefings, and numerous instructional articles for aviation publications. After returning to the United States in 1989, he worked as a pilot for Trump Air of Linden, New Jersey, flying Donald Trump's private AS332L Super Puma as well as the company's Sikorsky S-61s on regularly scheduled routes. Later, he worked as sales manager for Carson Helicopters of Perkasie, Pennsylvania.

Padfield's first book for McGraw-Hill, *Cross-Country Flying—3rd Edition*, was published in 1991. His second McGraw-Hill book, *Learning to Fly Helicopters*, was published in 1992. He also cowrote *To Fly Like a Bird*, the autobiography of Joe Mashman, which was published by the American Helicopter Society and *Rotor Wing International* magazine in 1992. (Mashman was one of the first and best-known test pilots with Bell Helicopter Corporation.)

Now residing with his wife and three children in Bethlehem, Pennsylvania, Randy Padfield is a full-time aviation writer and editor with *Aviation International News* of Midland Park, New Jersey. He also owns and flies a 1946 Taylorcraft BC-12D.

Other Related Titles of Interest

Aviator's Guide to GPS
Bill Clarke
A practical explanation of aviation's newest navigation system—the Global Positioning System: what it is, what it does, and how to use it.
0-07-011272-X $17.95 Paper
0-07-011271-1 $29.95 Hard

Avoiding Common Pilot Errors
John Stewart
Teaches students and licensed pilots the correct procedures for operating in controlled airspace. Covers preflight preparation, communications, phraseology and word concepts, regulations and procedures, and more.
0-07-155395-9 $17.95 Paper

The Pilot's Radio Communications Handbook, Revised 4th Edition
Paul E. Illman
Everything VFR pilots need to know to communicate from the cockpit effectively and use even the busiest airports with confidence. Now updated and expanded to cover new U. S. airspace designations.
0-07-031756-9 $17.95 Paper
0-07-031757-7 $29.95 Hard

Avoiding Mid-Air Collisions
Shari Stanford Krause, Ph.D.
Concise, easy-to-understand information on how to steer clear of other aircraft during all phases of flight. A virtual training course—and a unique, integrated approach to a serious safety issue.
0-07-035945-8 $16.95 Paper
0-07-035944-X $27.95 Hard

Pilot's Guide To Weather Reports
Terry T. Lankford
How to obtain and understand FAA weather reports, use current and forecasted weather to plan the best flight route, and decide whether to put off the flight to another day.
0-07-156003-3 $19.95 Paper

Stalls & Spins

Paul Craig

A practical guide to help private pilots and flight students meet new FAA requirements, this book demystifies stalls and spins. Readers gain a better knowledge and understanding of the aerodynamic principles involved, the psychological effects of stalling and spinning, the actions necessary to avoid disaster, and the spin characteristics of specific aircraft.

0-07-013422-7 $18.95 Paper
0-07-013421-9 $26.95 Hard